For EU product safety concerns:
IN TECH d.o.o., Prolaz Marije Krucifikse Kozulić 3, 51000 Rijeka, Croatia,
info@intechopen.com or visit our website at intechopen.com.

IntechOpen Book Series
Civil Engineering
Volume 10

Aims and Scope of the Series

Civil engineering is a traditional field of engineering from which most other branches of engineering have evolved. It comprises traditional sub-areas like transportation, structures, construction, geotechnics, water resources, and building materials. It also encompasses sustainability, risk, environment, and other concepts at its core. Historically, developments in civil engineering included traditional aspects of architecture and urban planning as well as practical applications from the construction industry. Most recently, many elements evolved from other fields of knowledge and topics like simulation, optimization, and decision science have been researched and applied to increase and evolve concepts and applications in this field. Civil engineering has evolved in the last years due to the demands of society in terms of the quality of its products, modern applications, official requirements, and cost and schedule restrictions. This series addresses real-life problems and applications of civil engineering and presents recent, cutting-edge research as well as traditional knowledge along with real-world examples of developments in the field.

Meet the Series Editor

Professor Assed N. Haddad is a Civil Engineer with a degree from the Federal University of Rio de Janeiro (UFRJ) earned in 1986, as well as a Juris Doctor degree from the Fluminense University Center earned in 1993, and a Master's degree in Civil Engineering from the Fluminense Federal University (UFF) obtained in 1992. He completed his Ph.D. in Production Engineering from COPPE / Federal University of Rio de Janeiro in 1996. Professor Haddad's academic pursuits have taken him to postdoctoral stays at the University of Florida, USA in 2006; at the Universitat Politècnica de Catalunya, Spain in 2010; and at the University of New South Wales Sydney, Australia in 2019. Currently, he serves as a Full Professor at the Federal University of Rio de Janeiro. He has held visiting professorships at various institutions including the University of Florida, Universitat Politècnica de Catalunya, Universitat Rovira i Virgili, and Western Sydney University. His research expertise encompasses Civil, Environmental, and Production Engineering, with a primary focus on the following topics: Construction Engineering and Management, Risk Management, and Life Cycle Assessment. He has been the recipient of research grants from the State of Rio de Janeiro, Brazil: CNE FAPERJ from 2019 to 2022 and from 2023 to 2025. Additionally, his research grants obtained from the Brazilian Government CNP since 2012 last to this date. Professor Haddad has been involved in several academic endeavors, being the Guest Editor of the International Journal of Construction Management; MDPI's Sustainability, Energies, and Infrastructures; Associate Editor at Frontiers in Built Environment / Sustainable Design and Construction; Guest Editor at Frontiers in Built Environment / Construction Management; and Academic Editor of the Journal of Engineering, Civil Engineering Section of Hindawi. He is currently a Professor of the Environmental Engineering Program at UFRJ and the Civil Engineering Program at UFF.

Meet the Volume Editor

Salvatore Antonio Biancardo is an associate professor at the Department of Civil, Construction and Environmental Engineering of the University of Naples Federico II. He is the co-founder of the University Spin-Off VIASTRATA for Infrastructure Building Information Management—infraBIM. In 2023, he earned the National Scientific Qualification to full professor. He served as a scientific committee member and chairman of many international conferences. He is an editorial board member/reviewer for several international indexed journals. His main fields of research are BIM for infrastructures, construction, and management of infrastructures.

Contents

Preface

Highway engineering is a multidisciplinary field with interconnected subdisciplines that include planning, safety, operations, design, and related fields such as structural, hydraulic, and geotechnical engineering. This book presents an up-to-date overview of practical knowledge to manage contemporary highways.

Section 1 – Policies Overview presents an overview of the following:

- Impacts that different types of best-value award algorithms have in the selection of design-builders in highway projects;
- Urban logistics guidelines for the efficiency of goods distribution in urban centers; and
- Adoption of a service load based on the characteristics of the vehicles on European roads.

Section 2 – Modeling, Construction, and Materials presents an overview of the following:

- Use of building information modeling as a structured and innovative methodology for designing, modeling, and managing transport and hydraulic infrastructures leveraging BIM;
- Opportunities and challenges of adopting 3D printing particularly to construct highway bridges;
- Performance evaluation of low-volume flexible pavement; and
- Risk matrices as environmental management tools.

Section 3 – Traffic Control and Safety presents an overview of the following:

- Traffic environment evaluation of "loopholes" using ETC2.0 data and geographic information systems;
- Implementation of a reinforcement learning algorithm to facilitate efficient travel for vehicles utilizing the fast lanes;
- Lane-keeping system integrating information from a road map, satellite receiver, and inertial sensors; and
- Road safety predictive analysis using machine learning algorithms for accident severity.

This professional book as a credible source and a valuable reference can be very applicable and useful for all professors, researchers, engineers, practicing professionals, trainee practitioners, students, and others interested in highway projects.

Salvatore Antonio Biancardo
VIASTRATA spin-off,
University of Napoli Federico II,
Napoli, Italy

Section 1

Policies Overview

Chapter 1

Best-Value in the Procurement of Highway Projects: Lessons Learned from US Design-Build Projects

Maria Calahorra-Jimenez and Gustavo Garcia-Melero

Abstract

Best value has been used for more than two decades in the procurement of design-build highway projects in the United States. The question of whether this type of procurement selects best-value proposers has been raised by several authors. Considering procurement data from 128 projects procured between 2000 and 2022, this chapter aims to provide insights on the impact that different types of best-value award algorithms have in the selection of design-builders as well as answers to the question of whether or not best-value procurement is actually selecting the best-value proposers. Ultimately, the chapter will reflect on lessons learned that might help highway administrators plan to use or use the best value in procuring highway projects.

Keywords: best-value procurement, design-build, highways, award algorithms, montecarlo simulation

1. Introduction

The design and construction of highway projects have been traditionally delivered using design-bid-build (DBB) and low-bid procurement. In contrast with the traditional low-bid procurement, best-value enables Departments of Transportation (DOTs) to consider non-cost criteria (e.g., technical qualifications), in addition to cost, to evaluate projects' proposals [1]. Best-value procurement in design-build (DB) highway projects is "a procurement process where price and other key factors are considered in the evaluation and selection of design-builders to enhance performance and value of construction" [2]. Best-value procurement has been used to select the firms that develop design-build highway projects in the United States [3]. This procurement enables Departments of Transportation (DOTs) to select a proposal by assessing cost and non-cost evaluation criteria [4]. Other criteria besides cost are incorporated to make a better selection that aligns with each DOT's objectives; given that selecting based only on price is one of "the greatest barriers for improvement" [5, 6].

The formula used to aggregate cost and technical criteria is usually named the award algorithm. According to Molenaar and Tran [7], in highway projects, the award algorithms most commonly used in best-value procurements are three: weighted

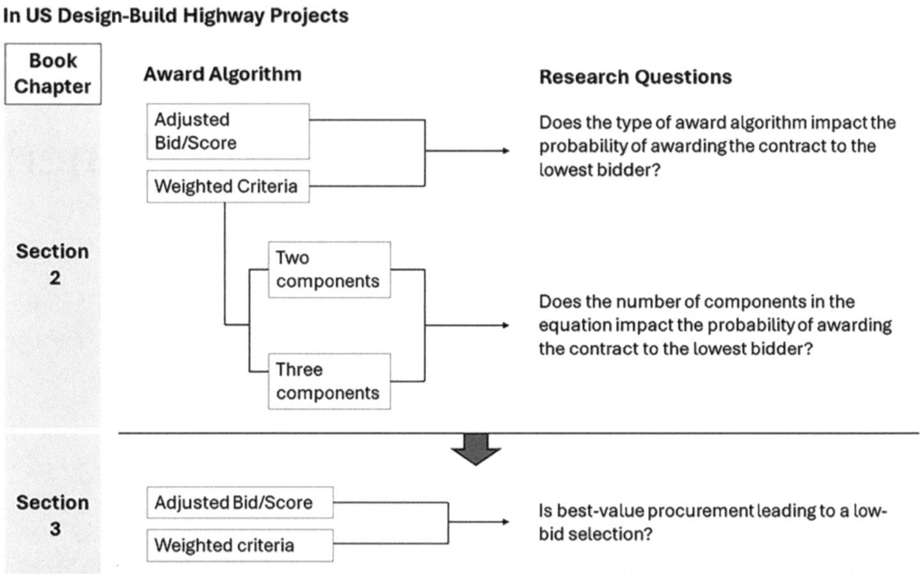

Figure 1.
Chapter structure.

criteria, adjusted bid, and adjusted score. In all of them, proposers submit a technical and a price proposal.

Different award algorithms might impact expectations about the importance of cost and technical components in the design-builder's evaluation. For example, Ballesteros-Pérez et al. [8] found that award algorithms are one of the variables that influence bidders' competitiveness.

After two decades of using best-value procurement in US transportation projects, research has shown that 80% of the projects procured using best-value were awarded to the lowest bidder [9, 10]. Based on these findings, it might be argued that best-value leads to a low-bid selection—and not to a best-value selection [11].

As shown in **Figure 1**, in this chapter, the authors explore, based on the analysis of historical procurement data, how the type of award algorithm might influence the probability of awarding the contract to the lowest bidder (Section 2) and also whether or not best-value procurement is leading to a low-bid selection (Section 3).

2. How do best-value award algorithms influence the selection?

The authors posed two research questions to understand how the type of award algorithm might affect the probability of the lowest bidder winning the contracts. The first question is *does the type of award algorithm—weighted criteria or adjusted bid/score —impact the probability of awarding the contract to the lowest bidder? And* the second question *does the number of components—two or three—in the weighted criteria algorithm impact the probability of awarding the contract to the lowest bidder*?

To answer these questions, the authors followed a four-step approach. First, score data were collected from DOTs across the United States. Score data included the scores given to design-build proposals regarding cost, technical, and qualification components. The aim of collecting these data was to build (in step 2) empirical probabilistic distributions for cost, technical, and qualification scores. Second, exploratory and goodness of fit analyses were conducted. Third, the Monte Carlo simulation was conducted using the probabilistic distributions obtained in step 2. Finally, results from the different scenarios simulated are compared.

2.1 Data collection

Best-value procurement scores were obtained from 128 design-build projects across 13 states in the United States (**Table 1**). The projects were procured between 2002 and 2020. Procurement's scores are generally public information made available through Departments of Transportation's Websites or under request.

Each project's procurement might include between two and five bids. Each bid contains one score related to the cost criterion, *CostScore;* one score regarding one non-cost criterion-cost, *NonCostScore1*; and, in some cases, might include another non-cost criterion score, *NonCostScore2*. In this research, *NonCostScore1* relates to the score given to the technical criterion (i.e., score given to the technical component of the proposal), which might consider aspects such as "project delivery approach," "conceptual road plans," "innovation and added value," "Environmental impacts & public outreach," etc., depending on the DOT [12].

NonCostScore2 relates to the score given to the qualifications criterion (i.e., the score given to the qualifications component of the proposal), which is the score that proposers receive in the first phase of the best-value procurement. Some states, such as the South Carolina Department of Transportation (SCDOT), include the qualifications score in the weighted criteria award algorithm as an additional criterion besides technical aspects [12].

In this study, each score is an observation. Thus, a total of 406 observations were analyzed from the 128 projects collected. Scores from different procurements and criteria were harmonized to a scale ranging from zero to one, as was considered by Calahorra et al. [11].

State	# Projects
Georgia	7
Kentucky	7
Louisiana	5
Minnesota	20
Mississippi	12
North Carolina	25
South Carolina	14
Texas	3
Virginia	4
Washington	24
California	2
Connecticut	2
Ohio	3
Total	**128**

Table 1.
Number of projects analyzed per state.

2.2 Data analysis

2.2.1 Variables definition

The authors defined nine [9] variables establishing a difference between the technical scores given to the lowest bidder (lb) and the non-lowest bidders (nlb) (**Table 2**).

Disaggregating the data enables a more accurate representation of the proposers' behavior that will be captured in the probabilistic distributions created in the second step of the data analysis and, subsequently, used in the simulation.

CostScore for the lowest bidders will not have a probabilistic distribution because, in common practice, it always adopts the value of 1. This means that for the cost criterion, the lowest bidder always gets one [1] as the maximum score in this criterion [13].

An exploratory data analysis (EDA) was conducted to analyze the variables in aggregated and disaggregated terms. The authors obtained the variables' statistics, checked missing values, removed inconsistent values, and calculated the correlation between aggregated variables.

2.2.2 Goodness of fit analysis

Statistical analysis was developed to obtain the probabilistic distributions that best fit each of the variables defined in **Table 2**. The Akaike Information Criteria (AIC) technique was used to measure the sample's fitness with a set of hypothesized distributions (**Tables 10, 12** and **14**). The authors utilized the Gamlss R package [14] to determine the empirical distributions that best fit the data using the AIC technique. The analysis resulted in five probability distributions for each of the five variables: NonCostScore1_lb (**Figure 5, Table 14**), NonCostScore2_lb (**Table 9**), CostScore_nlb (**Figure 3, Table 11**), NonCostScore1_nlb (**Figure 3, Table 13**), and NonCostScore2_nlb (**Table 9**).

Scores Aggregated	Scores Disaggregated	
	Lowest bidders	**Non-lowest bidders**
CostScore	1	CostScore_nlb
NonCostScore 1	NonCostScore1_lb	NonCostScore1_nlb
NonCostScore2	NonCostScore2_lb	NonCostScore2_nlb

Table 2.
Research variables.

Subscenarios of cost	The non-lowest bidder's cost proposal *[CostScore_nlb]* is
1	5% higher than the lowest bidder
2	10% higher than the lowest bidder
3	15% higher than the lowest bidder
4	20% higher than the lowest bidder
5	25% higher than the lowest bidder

Table 3.
Sub-scenarios for cost Score_nlb.

The simulation process in the next step used these probabilistic distributions as input for the analysis of scenarios.

2.3 Simulation

First, weighted criteria and adjusted/bid-adjusted score award algorithms were simulated. Second, the case of the weighted criteria award algorithm with two and three components was explored.

2.3.1 Weighted criteria and adjusted bid/score award algorithms

This section aims to determine whether using the weighted criteria or the adjusted bid algorithm influences the probability of awarding the contract to the lowest bidder. To this end, the authors created a theoretical procurement process with two bidders. One bidder was the lowest, and the other was the non-lowest bidder.

This situation was analyzed in two different cases. First, of the weighted criteria award algorithm and, second, the case of the adjusted bid.

Case 1: Weighted criteria.

In this case, the best-value weighted criteria award algorithm is defined with two components, one for cost and the other a non-cost criterion. The final score (*FS*) for each proposer is the result of the award algorithm formula in each case, as indicated by Eqs. (1) and (2).

$$FS\_lb = \mathrm{Wc}^{*}1 + \mathrm{Wnc1}^{*}NonCostScore1\_lb \quad (1)$$

$$FS\_nlb = \mathrm{Wc}^{*}CostScore\_nlb + \mathrm{Wnc1}^{*}NonCostScore1\_nlb \quad (2)$$

Where:

FS_lb is the final score for the lowest bidder.

FS_nlb is the final score for the no-lowest bidder.

NonCostScore1_lb, CostScore_nlb, NonCostScore1_nlb are the scores' probability distributions obtained from the goodness of fit analysis.

Wc and Wnc are the weights given to each criterion. The weights are fixed for each scenario simulation. In each case, four scenarios were simulated, varying the weights of cost between 70%, 60%, 50%, and 40%.

In each weights scenario, five sub-scenarios were created to allow the comparison with the adjusted bid case. Each sub-scenario represents a different relationship between the cost component of the lowest bidder and the non-lowest bidder, as indicated in **Table 3**.

The simulation was coded in R language and utilized 10,000 final score observations in each weight scenario to calculate the probability of the lowest bidder winning using Eq. 3.

$$\mathrm{Prob}\,(FSlb - FSnlb) > 0 \quad (3)$$

Case 2 Adjusted bid award algorithm.

In this case, the adjusted bid award algorithm considers two components, one for cost and the other a non-cost criterion. The final score (*FS*) for each proposer is the result of the award algorithm formula in each case, as indicated by Eqs. (4) and (5).

$$FS\_lb = 1/NonCostScore1\_lb \tag{4}$$

$$FS\_nlb = Cost\_nlb/NonCostScore1\_nlb \tag{5}$$

Where:

FS_lb is the final score for the lowest bidder.

FS_nlb is the final score for the no-lowest bidder.

NonCostScore1_lb, NonCostScore1_nlb are the scores' probability distributions obtained from the goodness of fit analysis.

CostScore_nlb is by the scenarios listed in **Table 3**.

The simulation was coded in R language and utilized 10,000 final score observations in each sub-scenario of *CostScore_nlb* scenario to calculate the probability of the lowest bidder winning using Eq. 6.

$$\text{Prob}\,(FSnlb - FSlb) > 0 \tag{6}$$

2.3.2 Weighted criteria two components and three components

This section aims to determine whether including two non-cost criteria instead of one in the weighted criteria award algorithm influences the probability of awarding the contract to the lowest bidder. To this end, the authors created a theoretical procurement process with two bidders. One bidder was the lowest, and the other was the non-lowest bidder.

This situation was analyzed in two different cases. First, the case of two non-cost criteria, and second, the case of one non-cost criterion,

Case 1: Two non-cost criteria.

In this case, the best-value weighted criteria award algorithm considers three components, one for cost and the other two for two non-cost criteria. The final score (*FS*) for each proposer is the result of the award algorithm formula in each case, as indicated by Eqs. (7) and (8):

$$FS\_lb = \text{Wc}^*1 + \text{Wnc1}^*NonCostScore1\_lb + \text{Wnc2}^*NonCostScore2\_lb \tag{7}$$

$$FS\_nlb = \text{Wc}^*CostScore\_nlb + \text{Wnc1}^*NonCostScore1\_nlb + \text{Wnc2}^*NonCostScore2\_nlb \tag{8}$$

Where:

FS_lb is the final score for the lowest bidder.

FS_nlb is the final score for the no-lowest bidder.

NonCostScore1_lb, NonCostScore2_lb, CostScore_nlb, NonCostScore1_nlb, and *NonCostScore2_nlb* are the scores' probability distributions obtained from the goodness of fit analysis.

Wc, Wnc1, and Wnc2 are the weights given to each criterion. The weights are fixed for each scenario simulation. In each case, 14 scenarios were simulated, varying the weights according to the values included in **Table 4**.

In each weights scenario, the probability of the lowest bidder winning was calculated by using Eq. 9

$$\text{Prob}\,(FSlb - FSnlb) > 0 \tag{9}$$

Case 2: One non-cost criterion.

Weight Scenario	Case 1: Two non-cost criteria			Case 2: One non-cost criterion	
	Cost [Wc]	NonCost1 [Wnc1]	NonCost2 [Wnc2]	Cost [Wc]	NonCost1 [Wnc1]
1	40%	10%	50%	40%	60%
2	40%	20%	40%	40%	60%
3	40%	30%	30%	40%	60%
4	40%	40%	20%	40%	60%
5	40%	50%	10%	40%	60%
6	50%	10%	40%	50%	50%
7	50%	20%	30%	50%	50%
8	50%	30%	20%	50%	50%
9	50%	40%	10%	50%	50%
10	60%	10%	30%	60%	40%
11	60%	20%	20%	60%	40%
12	60%	30%	10%	60%	40%
13	70%	10%	20%	70%	30%
14	70%	20%	10%	70%	30%

Table 4.
Weight scenarios.

In this case, the best-value weighted criteria award algorithm considers two components, one for cost and the other a non-cost criterion. The final score (*FS*) for each proposer is the result of the award algorithm formula in each case, as indicated by Eqs. (10) and (11).

$$FS\_lb = \mathrm{Wc}^{*}1 + \mathrm{Wnc1}^{*}\,NonCostScore1\_lb \tag{10}$$

$$FS\_nlb = \mathrm{Wc}^{*}\,CostScore\_nlb + \mathrm{Wnc1}^{*}\,NonCostScore1\_nlb \tag{11}$$

Where:

FS_lb is the final score for the lowest bidder.

FS_nlb is the final score for the no-lowest bidder.

NonCostScore1_lb, CostScore_nlb, and *NonCostScore1_nlb* are the scores' probability distributions obtained from the goodness of fit analysis.

Wc and Wnc are the weights given to each criterion. The weights are fixed for each scenario simulation. In each case, 14 scenarios were simulated, varying the weights according to the values included in **Table 4**.

The simulation was coded in R language and utilized 10,000 final score observations in each weight scenario to calculate the probability of the lowest bidder winning using Eq. 9.

2.4 Results

Results from the simulations run considering best-value procurements with weighted criteria, adjusted bid/score, and weighted criteria with two and three components' algorithms are summarized in the sections below.

Table 5 summarizes the results from the Montecarlo simulation related to the weighted criteria and adjusted bid/score award algorithms case.

Table 5 shows that the weighted criteria award algorithm leads to a lower probability of the lowest bidder winning if the weight of cost is 40%. In the rest of the scenarios of cost, the probability for different sub-scenarios of cost leads the adjusted bid algorithm to provide lower probabilities for the lowest bidder to win.

Table 6 summarizes the results from the Monte Carlo simulation related to the weighted criteria with two- and three-component cases.

Table 6 shows that the case of considering two non-cost criteria increases the probability of awarding the contract to the lowest bidder in all the scenarios. The highest increase is 21.48% in the case of 40% weight to the cost criterion and 60% to the *noncost2* criterion. As the weight of cost increases and the weight of the *noncost2* criterion decreases, the difference between the probability of awarding the contract to the lowest bidder in case 1 and case 2 is reduced up to a minimum of 2.78%.

Scenarios		**Probability lowest bidder to win**	
Weight of cost	**Cost of no-low bidder regarding the lower bidder**	**Case 1: Weighted Criteria Algorithm**	**Case 2: Adjusted bid algorithm**
40%	+5%	66.67%	69.10%
	+10%	75.60%	78.90%
	+15%	80.20%	83.60%
	+20%	87.00%	89.20%
	+25%	92.50%	93.60%
50%	+5%	70.40%	69.10%
	+10%	81.30%	78.90%
	+15%	89.40%	83.60%
	+20%	94.50%	89.20%
	+25%	97.10%	93.60%
60%	+5%	76.50%	69.10%
	+10%	89.50%	78.90%
	+15%	95.30%	83.60%
	+20%	98.00%	89.20%
	+25%	99.30%	93.60%
70%	+5%	84.90%	69.10%
	+10%	95.70%	78.90%
	+15%	98.80%	83.60%
	+20%	99.60%	89.20%
	+25%	100.00%	93.60%

Table 5.
Simulation results.

Scenario	Case 1: Two non-cost criteria [Equations 1,2]				Case 2: One non-cost criterion [Equations 4,5]			Variation between Case 1 and Case 2 [PC1]-[PC2]/ [PC2]
	Weights			Probability lowest bidder to win [PC1]	Weights		Probability lowest bidder to win [PC1]	
	Cost	Non Cost1	Non Cost2		Cost	Non Cost1		
1	40%	10%	50%	88.50%	40%	60%	72.85%	21.48%
2	40%	20%	40%	87.30%	40%	60%	72.85%	19.84%
3	40%	30%	30%	83.40%	40%	60%	72.85%	14.48%
4	40%	40%	20%	81.40%	40%	60%	72.85%	11.74%
5	40%	50%	10%	77.70%	40%	60%	72.85%	6.66%
6	50%	10%	40%	93.70%	50%	50%	78.40%	19.52%
7	50%	20%	30%	90.60%	50%	50%	78.40%	15.56%
8	50%	30%	20%	86.30%	50%	50%	78.40%	10.08%
9	50%	40%	10%	82.60%	50%	50%	78.40%	5.36%
10	60%	10%	30%	97.70%	60%	40%	85.07%	14.85%
11	60%	20%	20%	93.80%	60%	40%	85.07%	10.26%
12	60%	30%	10%	89.90%	60%	40%	85.07%	5.68%
13	70%	10%	20%	99.50%	70%	30%	93.60%	6.30%
14	70%	20%	10%	96.20%	70%	30%	93.60%	2.78%

Table 6.
Simulation results 2.

2.5 Discussion

This section addresses the questions posed at the beginning of the chapter.

2.5.1 Does the type of award algorithm impact the probability of awarding the contract to the lowest bidder?

DOTs aiming to use a weighted criteria algorithm to integrate cost and non-cost evaluation criteria are suggested to use weights for a cost of 50% or lower. If not, the effort that this type of evaluation requires is not worth it. The use of weighted criteria in these cases does not make a difference regarding the impact of non-cost criteria in the evaluation.

2.5.2 Does the number of components in the equation impact the probability of awarding the contract to the lowest bidder?

This study aimed to determine whether including two non-cost criteria instead of one in the weighted criteria award algorithm influences the probability of awarding the contract to the lowest bidder. Results from the Montecarlo simulation indicated that including a second non-cost criterion varies the probability of awarding the contract to the lowest bidder between 21.48% and 2.78%, depending on the weights given to the different criteria. This result suggests that the number of non-cost criteria has an influence on the likelihood of the lowest bidder being awarded the contract, as it does the ranges of the weights and scores ranges applied to those criteria, according to previous research [11].

Having around 80% of the best-value highway projects awarded to the lowest bidder is a reason to explore the impact different settings in the award algorithm might have on the likelihood of having the lowest bidder awarded the contract. Specifically, for the number of non-cost criteria to include in the weighted criteria formula, this research found that if the weight given to cost is lower than 50%, including two non-cost evaluation criteria instead of one increases the probability of awarding the contract to the lowest bidder in around 20.5%.

Further, Pomerol and Barbar-Romero [15] indicated that the weighted sum algorithm implies that one unit lost in one criterion is exactly compensated by one unit gained in the other. Thus, having multiple non-cost criteria might make it difficult for decision-makers to evaluate the trade-off that they are getting between cost and non-cost criteria.

Notwithstanding these results, DOTs might select several non-cost criteria in the award algorithm formula to obtain information about the proposers in all those areas. This research adds new data for decision-makers to consider when deciding how many criteria to include in the award algorithm.

As suggested by this research results, the number of criteria considered in the weighted criteria award algorithm influences the probability of awarding the contract to the lowest bidder. However, it is important to consider that these results depend on probabilistic distributions (for cost and non-cost criteria) obtained from historical data. Future research should explore what might be the results if the score distributions are different. Further, if it is found that the use of other distributions decreases the probability of the lowest bidder winning, future research should address what steps might be taken in actual practice to move the general trend to those distributions.

This research's findings contribute to the knowledge of alternative project delivery and procurement methods by showing how the number of criteria considered in the weighted criteria award algorithm influences the probability of awarding the contract to the lowest bidder. These results fill the gap existing in current practice regarding recommendations on the number of non-cost criteria to consider in the weighted criteria award algorithm, being helpful for DOTs crafting DB requests for proposals for highway construction projects.

3. Is best-value leading to a low-bid selection?

The authors posed one research question to explore whether best-value procurement is leading to a low-bid selection.

To answer this question, the research followed a three-step approach—first, data collection; second, best-value selection categorization; and third, descriptive statistical analysis.

3.1 Data collected

The authors collected data from design-build best-value bidding results from 275 transportation projects. The projects ranged from 2002 to 2021 and included 14 states across the United States. Best-value bidding results are information publicly available on DOTs' websites and include the following data for each project: state, year, project name, proposed cost, and technical scores given to the proposal placed in the project's procurement, and who was the winner in each case. From the data collected, the researchers reviewed and filtered all the projects with the objective of selecting only the ones that had complete information about the procurement and the selection decision. As a result, 113 projects were considered. Of those, 20% were procured using a weighted criteria award algorithm, and 80% adjusted bid/adjusted score.

3.2 Best-value selection type categorization

To characterize the type of selection that was conducted in each procurement, the researchers defined three categories: *Cost-driven selection*: the proposer that won the contract was the lowest bidder, and their technical proposal was not given the highest technical score among the proposers in that procurement. *Best-value selection*: the proposer that won the contract was the lowest bidder, and their technical proposal was given the highest technical score among the proposers in that procurement. *Quality-driven selection*: the proposer that won the contract was not the lowest bidder, and their technical proposal was given the highest technical score among all the proposers in that procurement.

3.3 Descriptive statistical analysis

The descriptive statistical analysis considered two nominal variables: the type of award algorithm that could take two values *weighted criteria* or *adjusted bid/score*, and the type of selection that could take three values: *cost-driven*, *best-value-driven*, or *quality-driven*. Cross-tabulation tables were used to understand the relationship between types of awards algorithms and types of selection.

3.4 Findings and discussion

Eighty-one percent of the 113 projects analyzed were awarded to the lowest bidder in either a cost-driven selection or best-value selection. Sixty-three percent of the projects were awarded to the firm that provided the best technical proposal.

Further, **Figure 2** shows two main trends:

The percentage of cost-driven or best-value-driven selections is similar when compared between award algorithms. In weighted criteria, the percentage of cost-driven and best-value-driven selections is 9%. In adjusted bid/adjusted score, the percentage ranges between 29% and 34%. These values are substantially higher than the ones obtained for quality-driven selections (3% and 17%, depending on whether the weighted criteria or the adjusted bid/score were used).

The weighted criteria algorithm provides similar percentages of cost-driven and best-value-driven selections (9%). However, the adjusted bid/adjusted score algorithms provide a higher percentage of best-value selections as compared with cost-driven selections (34% versus 29%).

Previous research has shown that after two decades of using best-value procurement, 80% of the projects were awarded to the lowest bidder [9, 10]. Based on these data, it might be argued that best-value procurement works as the low-bid procurement with a selection only based on cost.

The analysis of 113 best-value design-build projects supports previous research on the fact that around 80% of the projects were awarded to the lowest bidder (in this

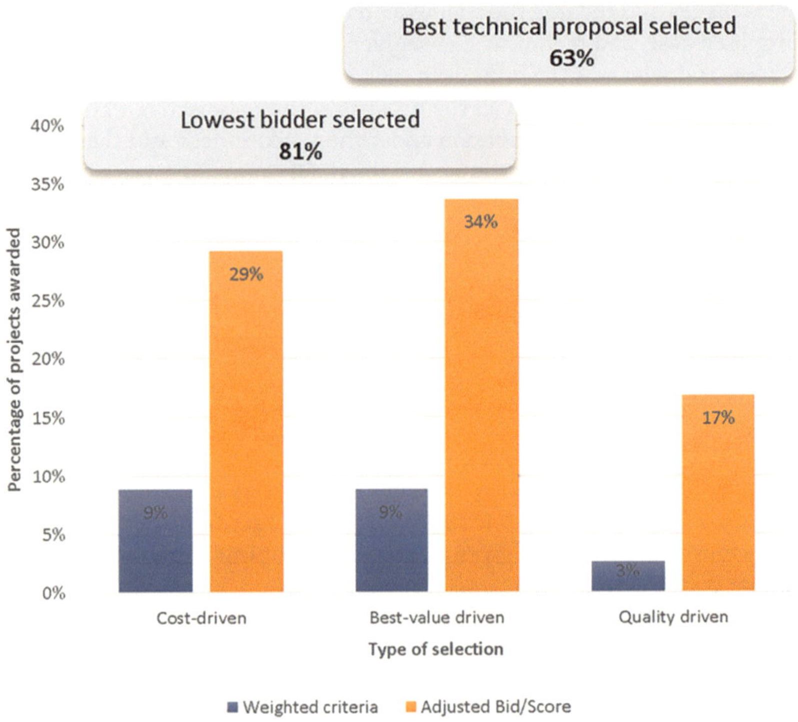

Figure 2.
Type of selection per type of best-value award algorithm. Note: Percentages calculated from the total sample of 113 projects.

research, the percentage was 81%). However, this research refutes the argument that best-value works as the low-bid procurement, considering that based on the results.

Independently of the award algorithm, less than 30% of the projects procured with best-value procurement provide a cost-driven selection.

When using the weighted criteria award algorithm, in 50% of the projects where the lowest bidder was selected, that firm was also the one that received the highest score in its technical proposal.

In the case of the adjusted bid/score algorithm, 53% of the projects awarded to the lowest bidder were awarded to the firms that received the highest technical score.

The aggregate percentage (weighted criteria plus adjusted bid/score) of cost-driven selections versus best-value selection and quality-driven selection together results in 38% versus 62%.

4. Conclusions

Results of these analyses show that the weighted criteria award algorithm with two components in the equation provides lower probabilities for the lowest bidder to win the contract as compared with the adjusted bid or weighted criteria with three components algorithms. It is important to consider that this applies in the case that the weight given to the cost component is lower than 50%.

Further, the analysis shows that 43% of the best-value design-build projects led to an actual best-value selection with the best technical and economical proposal being chosen. Thus, results suggest that there is still room for improvement, and public administrators need to be cognizant of the effect that the type of award algorithms, the weights given to each criterion and the score systems utilized have an impact in selection.

Acknowledgements

The authors wish to acknowledge the state highway agency personnel who supported this research. The authors would like to thank the economic support and assigned time provided by California State University, Fresno, and specifically the Lyles College of Engineering.

A. Appendices

A.1 Statistics summary and probability distributions

This section includes a description of the data collected based on the exploratory data analysis (EDA) conducted and the results of the goodness of fit analysis that provided the probabilistic distributions that best fit the cost scores and non-cost scores used in the analysis (**Tables** 7 and **8**). Following the description and goodness of fit

	CostScore	NonCostScore1	NonCostScore2
n	369	406	18
Minimum	0.4068	0.1653	0.5600
First quartile	0.8425	0.7811	0.5782
Median	0.9341	0.8600	0.6145
Mean	0.9069	0.8199	0.6202
Third quartile	1	0.9195	0.6580
Maximum	1	1	0.7110

Table 7.
Aggregated variables. Statistics summary.

	CostScore	NonCostScore1		NonCostScore2	
	_nlb	_nlb	_lb	_nlb	_lb
n	254	254	115	10	5
Minimum	0.4068	0.1653	0.2475	0.5770	0.5620
First quartile	0.8013	0.7844	0.8035	0.6112	0.5630
Median	0.8844	0.8560	0.8768	0.6460	0.5740
Mean	0.8647	0.8188	0.8424	0.6369	0.5984
Third quartile	0.9376	0.9250	0.9225	0.6662	0.5820
Maximum	0.9997	0.9952	1	0.6800	0.7110

Table 8.
Dissaggregated variables: Statistical summary.

Aggregated	Disaggregated	
	Low bidders	Non-low bidders
CostScore	1	*CostScore_nlb* *Distribution*: Box-Cox Power Exponential *Parameters:* Mu:-0.116678 Sigma: −2.437 Nu: 5.5734 Tau: 2.0596
NonCostScore1	*NonCostScore1_lb* *Distribution*: Box-Cox Cole and Green *Parameters*: Mu:0.865307 Sigma: −2.2464 Nu: 4.8876	*NonCostScore1_nlb* *Parameters:* Generalized Gamma Parameters: Mu:-0.07610 Sigma: −2.6457 Nu: 39.29
NonCostScore2	*NonCostScore2_lb* *Distribution:* Triangular *Parameters*: Min: 0.5620 Max: 0.7110 Mode: 0.60	*NonCostScore2_nlb* *Distribution*: Triangular *Parameters*: Min: 0.5770 Max: 0.6800 Mode: 0.64

Table 9.
Probability distributions for the cost score and non-cost score variables.

analysis, this section shows the results of Monte Carlo simulation performed in cases 1 and 2 described in **Table 4**.

CostScore and *NonCostScore1* distributions were found left-skewed (median > mean).

In the case of the disaggregated variables, the statistical summary is shown in **Table 8**. *CostScore_lb* is not included in the table, given that it always adopts the value of 1.

CostScore_nlb, *NonCostScore1_nlb,* and *NonCostScore1_lb* distributions were found left-skewed (median > mean). The small sample of NonCostScore2 did not allow for creating an empirical distribution.

The goodness of fit analysis resulted in the probabilistic distributions that best fit the different variables (**Table 9**). AIC and fitted distribution values are included in this article's Appendix. In the case of the variable *NonCostScore2*, a triangular distribution was adopted, given the sample size. The distribution was truncated between 0 and 1. The authors adopted a triangular distribution because, according to Johnson [16], it can be considered as a proxy for the beta distribution, and the beta distribution has been suggested by Law and Kelton [17] as "a rough model in the absence of data."

A.2 Fitted distributions

A.2.1 Variable: *CostScore_nlb*

See **Tables 10** and **11**, **Figure 3**

Distribution	AIC
Box-Cox Power Exponential Original	−529
Box-Cox Power Exponential	−529
Generalized Gamma	−521
Generalized Beta Type 2	−519
Box-Cox Cole and Green Original	−509

Table 10.
Top 5 distributions.

Parameter	Estimate	Std. Error	t value	Pr(>|t|)
Mu	−0.116678	0.007052	−16.55	<2e-16
Sigma	−2.437	0.053	−45.98	<2e-16
Nu	5.5734	0.5386	10.35	<2e-16
Tau	2.0596	0.3273	6.293	1.39e-09

Table 11.
Fitted distribution.

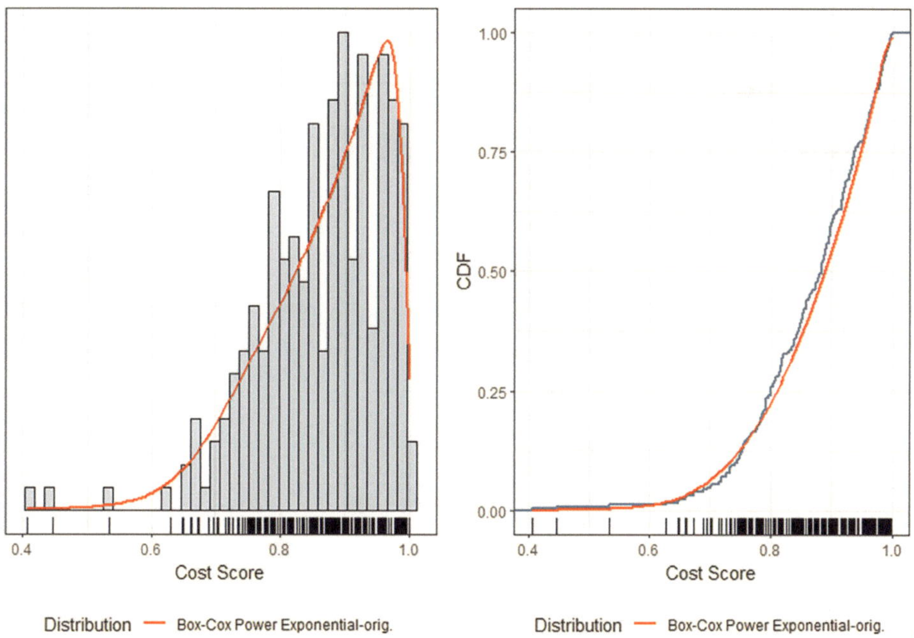

Figure 3.
Box-cox power exponential original distribution and CDF.

A.2.2 *NonCostScore1_nlb*

See **Tables 12** and **13**, **Figure 4**.

Distribution	AIC
Generalized Gamma	−383
Box-Cox Power Exponential	−381
Box-Cox Power Exponential Original	−381
Generalized Beta Type 2	−381
Box-Cox Cole and Green Original	−360

Table 12.
Top 5 distributions.

Parameter	Estimate	Std. Error	t value	Pr(>|t|)
Mu	−0.07610	0.01068	−7.127	1.09e-11
Sigma	−2.6457	0.1265	−20.92	<2e-16
Nu	39.29	10.44	3.764	0.000208

Table 13.
Fitted distribution.

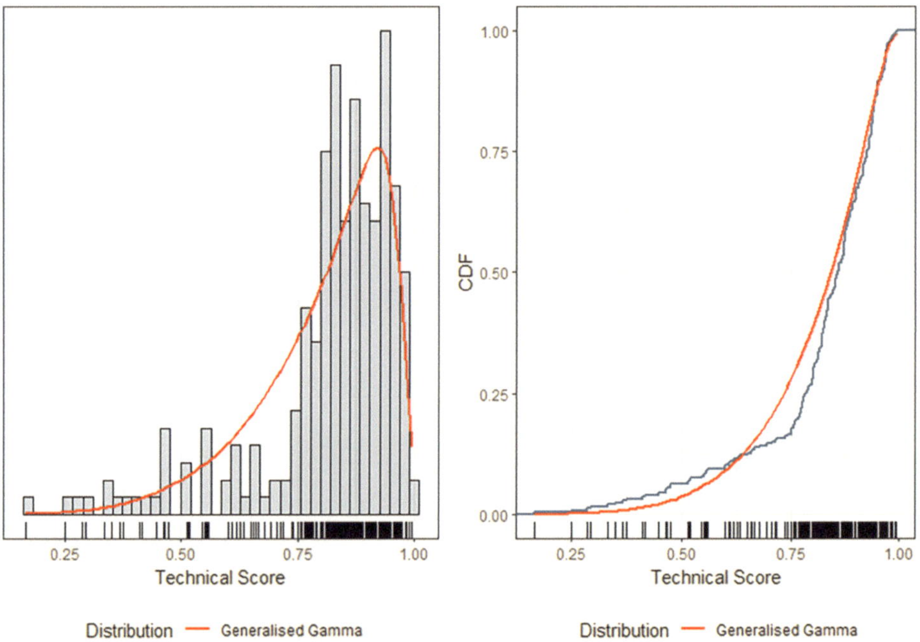

Figure 4.
Generalized gamma distribution and CDF.

A.2.3 *NonCostScore1_lb*

See **Tables 14** and **15**, **Figure 5**.

Distribution	AIC
Box-Cox Cole and Green	−197
Box-Cox Cole and Green Original	−197
Box-Cox Power Exponential	−195
Box-Cox Power Exponential Original	−195
Box-Cox t	−195

Table 14.
Top 5 distributions.

Parameter	Estimate	Std. Error	t value	Pr(>|t|)
Mu	0.865307	0.009517	90.92	<2e-16
Sigma	−2.2464	0.1114	−20.17	<2e-16
Nu	4.8876	0.7838	6.236	8.17e-09

Table 15.
Fitted distribution.

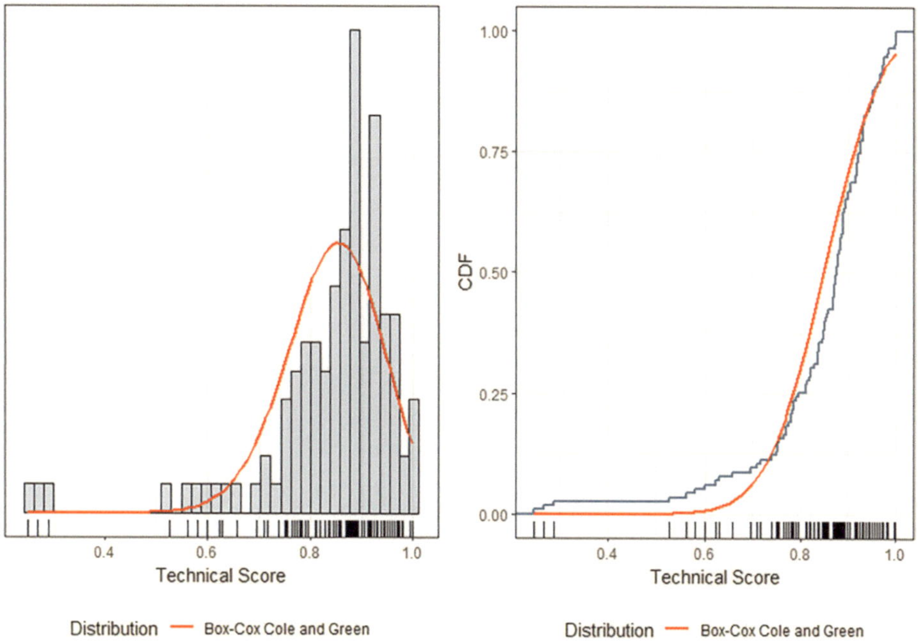

Figure 5.
Box-cox Cole and Green distribution and CDF.

Author details

Maria Calahorra-Jimenez[1*] and Gustavo Garcia-Melero[2]

1 California State University, Fresno, USA

2 University of Colorado, Denver, USA

*Address all correspondence to: mcalahorrajimenez@mail.fresnostate.edu

References

[1] Gransberg DD, Molenaar KR, Scott S, Smith N. Implementing best-value procurement in highway construction projects. In: Alternative Project Delivery, Procurement, and Contracting Methods for Highways. ASCE Press; 2007. pp. 60-79

[2] Scott S, Molenaar KR, Gransberg DD, Smith NC. NCHRP Report 561. Best-Value Procurement Methods for Highway Construction Projects. Washington D.C.: Transportation Research Board of the National Academies; 2006

[3] Anderson SD, Russell JS. NCHRP Report 451. Guidelines for Warranty, Multi-Parameter, and Best Value Contracting. Vol. 451. Washington D.C.: Transportation Research Board of the National Academies; 2001

[4] der Yu W, Wang KW. Best value or lowest bid? A quantitative perspective. Journal of Construction Engineering and Management. 2012;**138**(1):128-134

[5] Egan J. Rethinking the report of the construction task force. Construction. 1998;**38**:7

[6] Clement RT, Reilly T. Making Hard Decisions. Third ed. Mason, OH: Cengage Learning; 2014

[7] Keith M, Tran D. NCHRP Synthesis 471. Practices for Developing Transparent Best Value Selection Procedures. Washington D.C.: Transportation Research Board of the National Academies; 2015

[8] Ballesteros-Pérez P, Skitmore M, Pellicer E, Zhang X. Scoring rules and competitive behavior in best-value construction auctions. Journal of Construction Engineering and Management. 2016;**142**(9):04016035

[9] FMI (Fails Management Institute). Design-Build Utilization. Combined Market Study. 2018. Available from: https://dbia.org/wp-content/uploads/2018/06/Design-Build-Market-Research-FMI-2018.pdf

[10] Gaikwad SV, Calahorra-Jimenez M, Molenaar K, Torres-Machi C. Challenges in engineering estimates for best-value design-build projects: An analysis of bid dispersion in U.S. highway projects. Journal of Construction Engineering and Management. 2021;**147**(7):04021065

[11] Calahorra-Jimenez M, Torres-Machi C, Chamorro A, Alarcón LF, Molenaar K. Importance of noncost criteria weighing in best-value design–build US highway projects. Journal of Management in Engineering. 2021;**37**(4):1-12

[12] Molenaar K, Torres-Machi C. Efficiency Study of Design-Build Program [Internet]. 2020. Available from: https://rosap.ntl.bts.gov/view/dot/56640

[13] Scott S, Molenaar KR, Gransberg DD, Smith NC. Best-Value Procurement Methods for Highway Construction Projects. Best-Value Procurement Methods for Highway Construction Projects. Washington, D.C.: Transportation Research Board; 2006

[14] CRAN. Gamlss. Dist: Distributions for Generalized Additive Models for Location Scale and Shape. R programming language. Version 5.3–2. 2021

[15] Pomerol JC, Barba-Romero S. Multicriterion Decision in Management: Principles and Practice. New York: Springer; 2000

[16] Johnson D. The triangular distribution as a proxy for the beta distribution in risk analysis. The Journal of the Royal Statistical Society Series Statistical. 1997;**46**(3):387–398

[17] Law AM, Kelton WD. Simulation Modelling and Analysis. New York: MacGraw Hill; 1982

Chapter 2

Urban Logistics Guidelines for the Efficiency of Goods Distribution in Brazilian Cities

Antonio Gusmão

Abstract

The freight logistics operations within cities must be considered a relevant issue in city planning. The impacts caused by cargo transport on the urban distribution of goods, on the road and transport system in urban areas, as well as on interference with the road system deserve an appropriate analysis. The concepts of city logistics have become an efficient instrument to be adopted in the master planning of cities in order to provide an adequate logistical operation for loading and unloading goods in urban areas. It is also worth mentioning that current public policies practiced in the context of urban planning increasingly require guidelines that enable effective distribution of goods in cities. So, the objective of all the chapters is to present urban logistics guidelines for the efficiency of goods distribution in urban centers of Brazilian cities, taking the urban logistics concept as a reference.

Keywords: cargo distribution, congestion, city logistics, guidelines, urban mobility

1. Introduction

The urban freight logistics operation must be considered a relevant issue in city planning. Therefore, the impacts caused by road transport on the distribution of goods on the road and transport system in urban areas deserves an appropriate analysis, as there is still the problem of efficient distribution of cargo in various urban centers, especially in Brazil.

This research was developed based on a questionnaire administered to several experts in the field of transport.

The Brazilian gross domestic product (GDP) was 2.18 x 10^{12} U.S. dollars in 2023. GDP growth was 2.9% compared to 2022, when it was 1.98 x 10^{12} U.S. dollars [1].

Cargo transportation in urban centers has been occurring for a long time, and it has been verified that it is part of the growth of cities where social and economic activities take place. Thus, there is a greater increase in traffic accidents; the distribution and delivery of goods taking longer due to vehicle traffic congestion; and as a consequence there is an increase in different forms of pollution in the environment, which can be identified with road freight transport in cities as sound, vibration, air and environmental [2].

Thus, the concept of urban logistics proposed by Taniguchi [3] is an efficient tool to be applied in the development of cargo logistics operation planning for the distribution of goods in the urban area of cities.

This chapter contains seven items. After item 1, the introduction, item 2 presents knowledge of the urban transport system, with emphasis on freight movement in urban zones. Item 3 shows the city logistics conception and its definitions. Item 4 is key agents and their actions for the urban distribution of goods. Item 5 presents urban logistics guidelines. Item 6 shows proposed urban logistics guidelines and their benefits for the efficiency of urban distribution of goods. And finally, item 7 summarizes with the conclusions of the chapter.

2. Kowledge of urban transport system

The problem of cargo movement in urban areas should be seen as a specific issue due to the difficulties that the cargo sector encounters in its movement.

Browne et al. [4] and Prata et al. [5] highlighted that urban cargo transport has a fundamental importance in sustaining the population's lifestyle, as it plays an important role in the maintenance and conservation of industrial and commercial activities of cities.

According to *CNT* magazine [6], "a trade that does not receives its goods at the appropriate time for sale, when the customer wants to buy, brings direct reflections and consequences on your strategy service."

Therefore, it is necessary to try to propose urban load guidelines to make the supply and distribution of goods efficient, aiming at adequate planning of product movement in the cities.

2.1 Freight movement in urban zones

Cargo transport in urban areas can be considered a preponderant factor related to growth and the development of cities, being responsible for supplying goods necessary for the survival of the local population. In this sense, therefore, it is fundamental to quality of life.

Thus, it can be said that the demand for cargo transportation has in addition to other objectives the supply of cities, in order to provide them with their basic needs to allow the development of their activities.

There is, however, a fundamental need to supply goods in the urban centers of cities in order to guarantee the survival of local society and the carrying out of economic activities and social aspects of the city. On the other hand, the population is constantly changing and expanding, which causes the need to increase of the flow of goods to be distributed in urban centers.

Czerniak et al. [7] highlighted that "efficient planning of a cargo handling system must take into account the indicators economics of urban load." That is, good practices of key agents and participants in the process, such as shippers, transporters, and public power, involves understanding their perceptions and needs.

Road freight transport, therefore, has a especial importance when it is associated with regional development or growth of the economy of a medium or large city, as it is responsible for the supply and delivery of goods to cities that are part of the north–south and east–west connections of the country. With this type of transport, transporters must be concerned about the use of more appropriate fuels in their fleet for urban cargo operations, in order to have sustainable transport suited to the environmental conditions of cities and to promote quality of life.

Urban freight transport represents a sequence of activities within an economic process, be it local, regional, national, or global.

2.2 Main problems related to cargo movement in urban centres

According to data from the Ministry of Cities [8], the nine main metropolitan regions of the country concentrated 30% of the urban population urban at that time, and of this population, the majority is poor.

Later, with the IBGE census in 2010 [9], it was found that more than 80% of Brazilians lived in urban areas of the 5565 municipalities, with 29% of the total population being in 38 cities with more than 500,000 inhabitants, while 608 municipalities had population exceeding 50,000 inhabitants. The IBGE forecasts growth of the Brazilian population until 2042, when rates will become decreasing, with an increase in the urban population, especially in medium-sized cities [9].

Based on these data and knowing that the urban center of a city was defined by the Ministry of Cities as the place where economic activities take place— as it contains banks, shopping centers, supermarkets, and commerce in general, called the commercial region, where there are a range of jobs, types of business relationships (businesses), and also access to a wide range of goods and services —there is a significant increase in the circulation of cargo vehicles to adequately supply these centers. This circulation also increases travel times for road freight transport and, consequently, greater traffic congestion in cities.

Cargo movement in the main urban centers in cities of medium and large size (250,000 and 1 million inhabitants, respectively), especially in Brazil, which serve as a gateway to other locations, is carried out basically by road vehicles due to their ease of operation and existing infrastructure. This fact is confirmed by the impossibility of

Figure 1.
Traffic congestion in the direction of cities due to urban distribution of goods. Source: Globo/G1 (2022) [10].

service by other modes of transport directly from producer to consumer, in addition to being the only mode considered door-to-door.

Figure 1 presents a view of the related problem traffic congestion in cities, in relation to the circulation of freight vehicles on the road system, in a recent visual presented in 2022 [10]. It is also observed, at this time, that the problems already seen remain with the circulation of cargo vehicles in the function of distribution of goods.

Thus, the problems encountered in the transport system of main urban centers in the country can be compared to those of emerging countries such as Russia, India, and China, as described in the literature.

These problems refer not only to the accelerated growth of the fleet of motor vehicles that circulate in cities but also to the fleet of cargo transport vehicles which, due to the lack of appropriate policies to the distribution of goods, has greatly influenced and impacted the urban transport system [2].

2.3 Road cargo transport and its influence on urban centers of cities

The plan for the transit of cargo vehicles in Brazilian cities basically consists of specific regions with areas and corridors, depending on the capacity of each region to accommodate circulation and parking of vehicles, according to their dimensions and specific needs of each area. The ability to accommodate truck traffic is determined by the levels of concentration of activities and jobs and the saturation of the road system during peak hours. In many locations there are no suitable places for loading and unloading operations, as emphasized by Gusmão et al [11].

According to Allen et al. [12], in general, trucks use available or unavailable physical space such as curbs, park in double rows; block roads in the road system; cause congestion in traffic, and consequently interfere with the infrastructure of city roads. It is also verified that the operation of urban logistics may suffer economically and socially in its activities, as is seen throughout the day in the movement of urban cargo vehicles (VUCs) waiting for parking spaces or even the release of vacancies. So, it is necessary to provide urban logistics for efficiency of urban distribution of goods, as emphasized by Gusmão and Ribeiro [13].

Regarding Costa [14], in terms of impacts caused by the cargo transport in urban centers, the local public governments should control and check the activity of cargo logistics; however, the majority do not know how to do it. Thus, it can be assumed that whatever regulation is imposed by this public authority with the aim of ordering and guiding urban cargo transport (TUC) traffic in cities, the cargo will reach its destination in the right place and time, as a result of the process of logistical decision-making by the agents involved in the operation. Otherwise, it would be necessary to implement extremely strict access restrictions for the agents involved in the logistics operation to restructure their supplies, which could compromise the principles of freedom of movement and commercial activities.

Browne et al. argued [15] that urban cargo transport has different characteristics that can be changed in the long term to reduce their impacts along with related influences that occur as a result of there impacts.

3. City logistics conception and definitions

Some authors emphasize the city logistics conception, and they give their definitions as follows.

Taniguchi [16] says, "Urban Logistics, therefore, in short, refers to techniques and projects that, through the involvement of public and private agents, aim to reduce the total number of trips/vehicles in urban areas and/or reduce their negative impacts."

The concept of urban logistics, in the view of Dutra [17], arises as a response to the great need for organization of entities that work with the movement of goods within urban spaces because of the deterioration in the population's quality of life. This new concept raises concerns regarding the effectiveness and efficiency of operations and movements involving cargo transport.

According to the Report World Cities 2022, published by UN Habitat [18], the world population will be 68% urban by 2050, rising from the global total of 56% in 2021. According to the same report, 59 countries have more than 80% urban population. Despite a slowdown in the pace of urbanization during the pandemic, the urban population is expected to increase by 2.2 billion people annually by 2050.

In the view of Muñuzuri et al. [19], urban logistics is the term used to denote specific logistical operations involved in goods distribution in urban areas with their specific problems related to urban road traffic and its influence on the transport system, such as delays caused by congestion.

According to Oliveira [20], urban logistics aims to reduce diseconomies to make the entire system more effective through innovative solutions that reduce logistical problems generated by distribution in urban areas and improve quality of life.

According to Benjellou et al. [21], urban logistics is a concept that surrounds the domain of ideas, studies, policies, models, and methods that allow achieving the objectives of reducing congestion and increasing mobility through the control of VUCs that operate in urban areas, aiming for greater efficiency in freight circulation in these centers and reducing levels of environmental and noise pollution, aiming to contribute to the objectives established by the Kyoto Protocol [22], with the consequent improvement in the quality of life of the city's population.

Therefore, it is extremely important to propose urban logistics guidelines, based on the city logistics concept, to achieve efficient distribution and delivery of goods in the urban centers of large and medium-sized Brazilian cities, said Gusmão and Ribeiro [13].

4. Key agents and their actions for the urban distribution of goods

Taniguchi et al. [23] presented a framework for viewing the concept of urban logistics that was originally based on three fundamental pillars: sustainability, mobility, and quality of life.

Later, Teo et al. [24] brought a new vision with the pillars of mobility, sustainability, habitability, and resilience.

These pillars are related to values such as global competitiveness, efficiency, respect for the environment, reduction of congestion, trust, safety, energy conservation, and workforce strength.

Reducing congestion is one of the important values that relate to the four pillars of urban logistics. This indicates that it is a fundamental element that needs to be considered when thinking about reestablishing the balance of the road system and the use of the transport system in cities, taking as reference the adoption of specific guidelines based on the city logistics design. This should be a relevant issue to be observed by transport planners when designing urban transport systems.

Figure 2 presents an overview of the key agents involved in the urban distribution of goods (shippers, cargo transporters, consumers, and public authorities). This

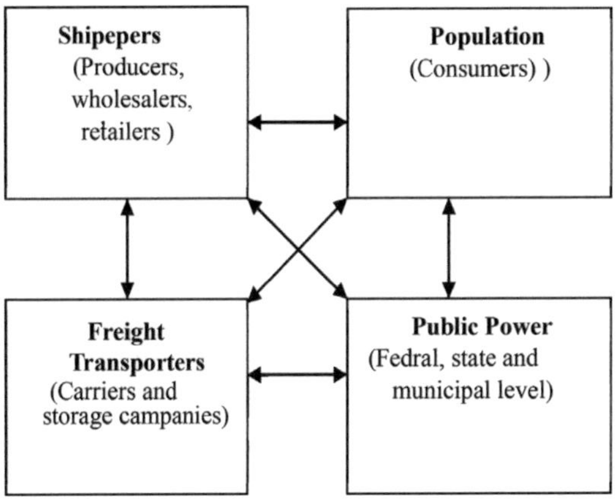

Figure 2.
Key agents in city logistics: Source: Taniguchi et al. [25].

representation establishes a partnership relationship between key agents within the market structure in which economic activities are the elements responsible for its support.

According to **Figure 2**, it is essential to have a partnership and full integration between the different key agents that participated in the vision of urban logistics in the distribution of goods. This allowed the desired efficiency to be achieved in logistical operations for loading and unloading goods in urban centers, aiming to carry out economic activities that supported the market, bringing direct benefits to society.

Therefore, it is essential to carry out research, so that important data can be obtained that, after due analysis, allows key agents to carry out specific actions using appopriate logistic operations that can lead to the desired efficiency of distribution and delivery of goods in the main Brazilian cities.

5. Urban logistics guidelines

This work sought to understand the problem of distributing goods in urban centers, which is normally carried out by road transport. With this knowledge, it was possible to identify the problems and impacts caused by road transport in the loading and unloading function of goods to supply cities. With this, it was possible to note that it is still necessary, in Brazil, to create a specific public policy that treats urban cargo logistics as a fundamental element of urban planning.

5.1 Development for proposing guidelines

The advancement of the study presents the city logistics/urban logistics concept as a new element of transport planning, capable of balancing the transport system of cities with the social costs involved, which has been used on a large scale in Europe since the end of the 1990s, with success in several cities aiming at the efficiency of urban distribution of goods.

The BESTUFS project was developed in Europe, initially between 2000 and 2003, with the aim of identifying and disseminating best practices in relation to urban freight transport. At the end of 2004, the first BESTUFS Project Report [26] was compiled, and the second BESTUFS Project at the end of 2007 [27], which served as a reference for new initiatives, adopting city logistics with the purpose of making the distribution of goods efficient in the main European urban centers, having been tested in some European cities such as Rome, Stuttgart, Paris, and others with positive and successful results regarding the efficiency of urban distribution of goods.

In Brazil, projects and experiences using the urban logistics concept were developed, after the analysis of Sanches Junior [28]. It is observed that it is necessary to propose urban logistics guidelines for planning cargo movement in cities, and that it constitutes a complement to the general guidelines established in PNMUS/2004 [29], and in Federal Law No. 12,587/2012 [30]. So, it is necessary to prepare and complete the urban mobility plan and master plan, in accordance with the reference notebook for preparing the urban mobility plan—PlanMob [31].

It is worth highlighting, therefore, that this work contributes to proposing guidelines for urban logistics at a time when Brazilian cities need to become efficient in the supply and distribution of urban cargo, through the adequate movement of goods in Brazilian urban centers, aiming to meet the purposes of urban mobility and sustainability of cities.

5.2 Understanding urban logistics guidelines

Urban logistics guidelines are understood as procedural standards that allow public managers to manage actions related to the supply and distribution of goods in urban centers, which include loading and unloading activities.

Urban cargo logistics, therefore, must contemplate the operability of the standards that are contained in the guidelines that must be part of the urban planning transport plan to promote the execution of the various economic activities that promote the development of cities.

Due to the gaps in the urban mobility plan (PlanMob) regarding specific guidelines related to urban load, in view of the nonspecification of such guidelines by the Ministry of Cities after the promulgation of Federal Law n° 12,587/2012 [30], which established the general guidelines of the National Urban Mobility Policy, it is necessary to establish guidelines within the master plans still under development in Brazilian municipalities. In this way, the proposed urban logistics guidelines are an opportunity to effectively establish conditions so that Brazilian urban centers can be efficient in the urban distribution of goods and thus contribute to an adequate and satisfactory quality of life for the population.

6. Proposed urban logistics guidelines

This study proposes urban logistics guidelines that allow the elaboration of urban planning within cities' master plans, highlighting urban mobility and its sustainable development.

Figure 3 presents the process developed to propose urban logistics guidelines.

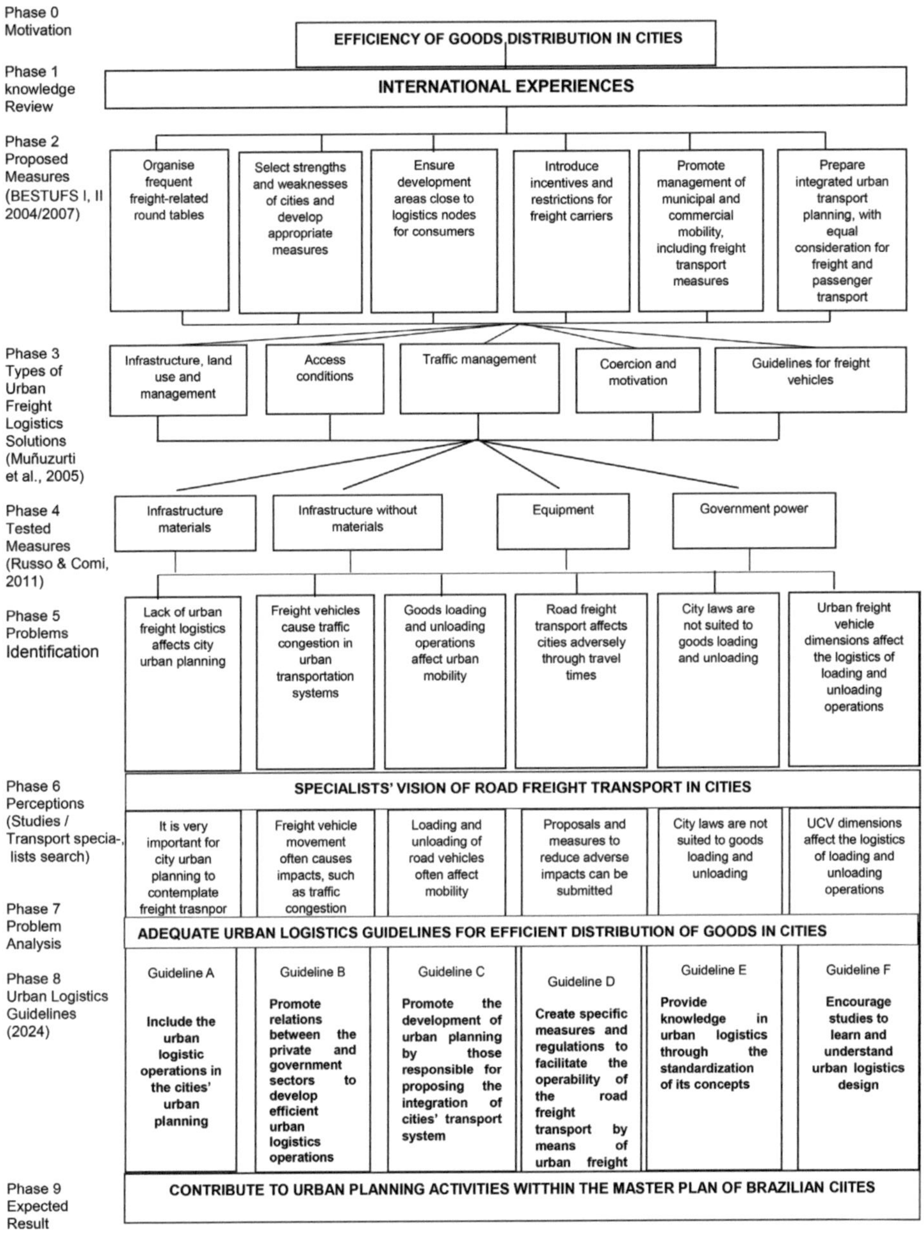

Figure 3.
Urban logistics guidelines for efficient distribution of goods in Brazilian cities.

6.1 Urban logistics guidelines

The urban logistics guidelines proposed here for the efficiency of goods distribution in Brazilian cities are based on the development of proposing guidelines shown in item 5.1 and the concept of city logistics emphasized by Taniguchi [32].

6.1.1 Guideline A: include the urban logistic operations in the cities' urban planning

To meet the fundamental necessity of goods distribution in the cities, it is necessary and desirable that urban logistic is definitively inserted into the urban planning as a way of establishing the balance of the city transport system, aimed at improving the quality of life in society.

For the cities that already have developed their urban mobility Plan and do not have specific guidelines for urban logistics, this is an opportunity to review and adjust their master plan so that they can meet current and future sustainability needs toward promoting city development.

Benefits

This guidelines will effectively contribute to:

- Improved performance of the urban logistics activities
- The equilibrium of the urban transport system in cities
- Meeting the needs of cargo distribution through urban transport to satisfy the requirements of the PNMU provided for in law 12,587/2012
- The urgency of having appropriate solutions regarding the various impacts caused by road freight transport in order to meet the needs of consumers in cities due to the lack of public policies for this purpose
- New research and studies seeking to reduce the negative impacts caused by road freight transport on the city's road system.
- Establishment of a modern urban transport system that provides an adequate quality of life for the population of cities as well as the implementation of various economic and social activities that allow the sustainability of cities
- Adoption of the city logistics concept with its appropriate concepts, in order to carry out urban logistics activities to promote urban mobility and the sustainable development of future cities
- Urban distribution of goods carried out efficiently, promoting the inclusion of logistical operations in the urban planning of cities, with emphasis on an efficient integrated transport system

6.1.2 Guideline B: promote relations between the private and government sectors to develop efficient urban logistics operations

The proposition of this guideline aims at the establishment of city logistics techniques that provide the integration between the private and government sectors with the involvement of key agents for an effective and successful implementation process.

Benefits

This guideline will effectively contribute to:

- The achievement of the urban logistics activities in cities

- The accurate and adequate planning of activities to be undertaken for the implementation of the city logistics technique in cities

- The adequacy and effective implementation of activities considered important for the management of urban logistics operations

- Elaboration of efficient urban planning, including transport systems that lead to effective application of PNMU guidelines

- The need to satisfy the wishes of society regarding solutions that seek the reduction of the impacts brought by road freight transport in cities, and that may lead to a condition of sustainability in the future cities

- The adequacy and improvement of the transport system in the establishment of conditions of use and benefit for population of the cities

- The possibility of applying the city logistics concepts, seeking to establish the condition of urban mobility and sustainable development for cities

- The implementation of the city logistics conception seeking as result the efficiency of goods distribution in the urban centers

6.1.3 Guideline C: promote the development of urban planning by those responsible for proposing the integration of cities' transport systems

Given the importance of the freight in the cities as a key element for the development and growth of their economic activities, the supply and distribution of goods is of vital importance for the survival and maintenance of such cities.

So, we propose the guideline of "Promote the development of urban planning by those responsible for proposing the integration of cities' transport systems."

Benefits

This guideline will help to:

- Encourage an adequate road system to promote traffic fluidity and allow adequate urban mobility for pedestrians and road freight transport vehicles in cities

- Minimize negative impacts and their consequences in relation to the cities' transport system that do not allow the population to have an adequate quality of life to achieve their objectives

- Encourage cities to have an integrated transport system, enabling lower costs for the population, allowing for appropriate urban mobility

- Allow the different types of cargo that are distributed and delivered in cities to have adequate treatment in accordance with the guidelines of the National Policy for Sustainable Urban Mobility

- Promote meeting the needs of the population of cities, seeking to obtain appropriate solutions that allow minimizing the different impacts caused by road freight transport in urban centers

- Encourage cities to have a more modern and updated urban transport system so that the needs of the population are met, contributing to the development of local activities and a better quality of life

- Encourage the adoption of the urban logistics concept as a way of promoting the efficiency of urban mobility in cities and their sustainable development

- Allow urban logistics actions to be adopted in the planning of the integrated urban transport system within the cities' master plans to obtain the efficiency of the distribution of goods in urban centers

6.1.4 *Guideline D: create specific measures and regulations to facilitate the operability of the road freight transport by means of urban freight vehicles*

It has been observed that city authorities believe that urban freight vehicle regulations must require the creation of a new cargo vehicle that has the following mandatory items: a clean engine, noise reduction technologies to VUC standards, more vehicle handling and maneuvering, and easier loading and unloading of goods.

Benefits

The adoption of new rules for urban vehicle load will allow the following:

- Standardize the types of vehicles that will carry out loading and unloading activities within the city's logistics operations, with emphasis on the movement of motorized cargo vehicles (trucks) with more than two axles, which lead to traffic congestion in cities, improving traffic flow on the road system

- Encourage the adoption of appropriate urban mobility in order to allow traffic fluidity, decongesting the city's road system with the loading and unloading operations of heavy cargo vehicles (trucks) and seeking to minimize the impacts caused by their movement in urban centers

- Seek to develop more effective and efficient control of the impacts caused by VUCs that circulate in urban centers

- Monitor the routes followed by VUCs in the distribution and delivery of goods

- Verify the benefits obtained and the quality-of-life conditions that improve depending on the different activities carried out by cities for the development of their enterprises

- Improve the quality of life of the city's population through an adequate urban transport system based on efficient regulation of the traffic system where VUCS circulate.

6.1.5 *Guideline E: provide knowledge in urban logistics through the standardization of its concepts*

The purpose of this guideline is to unify and standardize the knowledge of urban logistics and urban cargo, since there is an issue of lack of knowledge of such matters in various Brazilian geographic regions.

Benefits

This guideline will bring the following benefits:

- Standardization of knowledge of urban load for a greater understanding of the importance of the distribution of urban goods
- Increased efficiency of logistics operations in Brazilian cities with the aim of implementing urban logistics concepts effectively within the cities' master plan
- Properly practice urban logistics operations in the road freight transport system
- Carry out the distribution and delivery of goods in cities, with the aim of encouraging the development of their economic and social activities
- Encourage adequate treatment of the loading and unloading of goods in urban city centers, in accordance with the guidelines of the National Urban Mobility Policy
- Practice urban logistics design appropriately, with the aim of allowing urban mobility to be sustainable for cities as well as the practice of their activities

6.1.6 Guideline F: encourage studies to learn and understand urban logistics design

This guideline aims to encourage studies to learn and understand urban logistics design of cities. Further studies and research need to be conducted about urban cargo and its relationship, importance, and benefits to the city activities.

Benefits

This guideline will bring the following benefits:

- Understanding of what urban logistics is by everyone who works with this activity, with the involvement of all key agents who intend to put their concepts into practice in logistics operations in cities
- Practice the urban distribution of goods through efficient use of urban logistics design in Brazilian cities
- Develop adequate planning for the urban transport system within the city's master plan, based on the concepts of urban logistics, in order to allow efficient delivery of goods in urban centers
- Learn and understand more intensively what happens in the complexity of logistical operations used in loading and unloading goods within cities
- Know about urban freight transport as well as urban logistics and their effects on various activities in cities

7. Conclusions

This chapter showed urban logistics guidelines to improve the efficiency of goods distribution in the urban centers of Brazilian cities. The literature search on urban transport systems has identified the need to have the distribution of goods in urban centers carried out in a more appropriate and efficient way to provide greater benefits to the population of cities.

According to the literature consulted, it can be concluded that the practice of urban freight logistics in Brazil is being developed based on the negative externalities created in the urban transport system due to population growth, the demographic expansion of cities, and the demand for urban mobility and sustainable urban development.

The city logistics conception, which has been successfully used in several cities around the world, is a field of knowledge that aims to globally optimize logistics operations in urban areas and that can bring relevant benefits in its practice to Brazilian cities.

Current public policies adopted in urban planning within the cities' master plans require specific guidelines for urban logistics so that the efficiency of distribution and delivery of goods desired by society can be achieved.

The gap of several years that distances Brazil from Europe and more developed countries on other continents, regarding urban cargo logistics, indicates that appropriate solutions can be obtained for this issue. Administrators (public authorities in their different instances) must recognize the importance of this activity for the development of Brazilian cities, together with other participants who must be part of the process of urban logistics initiatives (city logistics), so that there is effectively adequate planning for the urban distribution of goods in Brazilian cities.

Conflict of interest

The authors declare no conflict of interest.

Author details

Antonio Gusmão
Federal Fluminense University, Niterói, Brazil

*Address all correspondence to: antoniogusmao@id.uff.br

References

[1] IBGE, Instituto Brasileiro de Geografia e Estatística (Brazilian Institute of Geography and Statistics). PIB Preço Corrente, Brasília, DF. 2020. Available from: https://www.brasilemsintetese.ibge.gov.br/contasnacionais/pib-valorescorrentes [Accessed: April 06, 2024]

[2] Gusmão A. Urban cargo guidelines for Brazilian urban centers based on city logistics [thesis]. Rio de Janeiro: Federal University of Rio de Janeirol; 2016

[3] Taniguchi E, Thompson RG, Yamada T, Duin RV. City Logistics—Network Modelling and Intelligent Transport Systems. Pergamon: Amsterdam; 2001. 252 p. DOI: 10.1108/9780585473840

[4] Browne M, Sweet M, Woodburn A, Allen J. Urban freight consolidation centers. In: Final Report. London: University of Westminster; 2005

[5] Prata B, Oliveira L, Dutra N, Pereira Neto W. Urban Logistics: Fundamentals and Applications. Curitiba: CRV. 288. p; 2012. ISBN: 978-85-8042-326-6

[6] CNT Magazine. Essential delivery. 2012. Available from: http://www.cnt.org.br; http://www.sestcenat.org.br [Accessed: April 10, 2012]

[7] Czerniak R, Lahsene J, Chatterjee A. Urban Freight Movement: What Form Will it Take? Committee on Urban Goods Movement: Transportation in the New Millennium. Transportation Research Board; 2000. A3B05: Committee on Safety Data, Analysis, and Evaluation Washington, DC United States; 2005. Accession Number: 00783807

[8] Ministry of Cicades. Integrated urban mobility management course. Brasília. 2006. Available from: http://www.cidades.gov.br [Accessed: October 18, 2001]

[9] Brazilian Institute of Geography and Statistics (IBGE). 2010 census. 2011. Available from: http://www.ibge.gov.br/cidadesat?deful.php [Accessed: November 28, 2011]

[10] GLOBO/G1. Too many trucks congest the conego Domenico highway Rangoni and the ferry crossing in Guarujá. 2022. Available from: https://g1.globo.com/sp/santosregiao/noticia/2022/03/16/excesso-de-caminhoes-congestiona-the-rodovia-conego-domenico-rangoni-and-the-ferry-crossing-in-guarujsp.ghtml [Accessed: July 20, 2023]

[11] Gusmão A, Santos M, Fraga C, Lopes CN, Amorim DP. The importance of regulation for the effectiveness of planning in different modes of transport. In: Paper Presented at the VIII Congress of the State of Rio de Janeiro on Transport, 7-8 August, Rio de Janeiro, 2010. 2010. Rio de Transportes. Available from: https: www.riodetransportes.org.br [Accessed: July 20, 2023]

[12] Allen J, Anderson S, Browne M, et al. A framework for considering policies to encourage sustainable urban freight traffic and goods/service flows. In: Report 3: Making Urban Goods and Service Operations more Sustainable: Policy Measures and Company Initiatives. London, England: University of Westminster; 2000

[13] Gusmão A, Ribeiro P. Guidelines for the efficiency of urban goods distribution: The Brazilian case. Case Studies on Transport Policy. 2020;**8**:1478-1488. DOI: 10.016/j.cstp.2020.10.013

[14] Costa E. Methodology for defining the sub-network for the transport of dangerous products in urban areas [thesis]. Rio de Janeiro: Federal University of Rio de Janeiro. 2009

[15] Browne M, Allen J, Nemoto T, et al. Reducing social and environmental impacts of urban freight transport: A review of some major cities. Procedia Social and Behavioral Sciences. 2012;**39**:19-33. DOI: 10.1016/j.sbspro.2012.03.088

[16] Taniguchi E. Rensselaer polytechnic institute and institute for city logistics. Short Course on City Logistics. Kyoto, Japan: Kyoto University; 2002

[17] DUTRA NGS. The Focus of "City Logistics" in Urban Parcel Distribution. [D.Sc. Thesis]. Florianópolis, SC, Brazil: Federal University of Santa Catarina; 2004

[18] United Nations (UN). Envisaging the future of cities. 2022. Available from: https://nacoesunidas.org/no-dia-mundial-das-cidades-onu-propoe-debate-acerca-dos-desafios-da-urbanizacao-global [Accessed: June 15, 2023]

[19] Muñuzuri J, Larrañeta J, Onieva L. et al. Solutions applicable by local administrations for urban logistics improvement. Cities. 2005;**22**(1):15-28

[20] Oliveira L. Modeling to evaluate the feasibility of implementing a small parcel distribution system within the concepts of city logistics [thesis]. Florianópolis: Federal University of Santa Catarina. 2007

[21] Benjelloun A, Crainic T. Trends, challenges, and perspectives in city logistics. Buletinul AGIR. 2009;**4**:45-51

[22] Kyoto Protocol. Constitution of the Kyoto protocol. 1997. Available from: https://antigo.mma.gov.br/clima/convencao-das-nacoes-unidas/kyoto-protocol.html [Accessed: October 23, 2023]

[23] Taniguchi E, Thompson R, Yamada T. Visions for city logistics. In: Logistics Systems for Sustainable Cities. Proceedings of the 3rd International Conference on City Logistics. Madeira, Portugal: Emerald Group Publishing Limited; 25-27 Jun 2003. pp. 1-16. DOI: 10.1108/9780080473222-001

[24] Teo J, Taniguchi E, Qureshi A. Evaluation of load factor control and urban freight road pricing joint schemes with multi-agent systems learning models. Procedia–Social and Behavioral Sciences. 2014;**125**:62-74. DOI: 10.1016/j.sbspro.2014.01.1456

[25] Taniguchi E, Thompson R, Yamada T, Duin R. City Logistics: Network Modelling and Intelligent Transport Systems. Wagon Lane: Emerald; 2007. 260 p. DOI: 10.1108/9780585473840. ISBN: 978-0-08-043903-7

[26] BESTUFS I. Best urban freight solutions project. 2004. Available from: http://www.bestufs.net/ [Accessed: May 20, 2020]

[27] BESTUFS II. Best urban freight solutions Project. 2007. Available from: http://www.bestufs.net/ [Accessed: July 20, 2020]

[28] Sanches Junior P. Urban logistics: An analysis of the Brazilian reality [thesis]. Campinas: State University of Campinas; 2008

[29] Brazil. National policy for sustainable urban mobility: Document for discussion. 2004. Available from: http://www.cidades.gov.br [Accessed: October 18, 2012]

[30] Brazil. Law No. 12,587 of 3 January 2012. Establishes the guidelines of the National Urban Mobility Policy. 2012. Available from: https://www.planalto.gov.br/ccivil_03/_ato2011-2014/2012/lei/l12587.htm [Accessed: June 01, 2012]

[31] National Secretariat for Transport and Urban Mobility – SeMob, Ministries of Cities, Brazil. Reference Booklet for Elaborating an Urban Mobility Plan—PlanMob. Ministry of Cities. 2015. Available from: https://itdpbrasil.org/planmob/: Brasília [Accessed: Septmber 20, 2015]

[32] Taniguchi E. Concept of city logistics. Institute for City Logistics. Kyoto University. In: Presentation at the XXV Congress of Research and Teaching in Transport, 11-14 November 2011. Belo Horizonte, Minas Gerais, Brazil; 2011

Chapter 3

On the Need to Increase the Design Load and Adoption of a Uniform Service Load for Bridges in Europe

Janusz Rymsza

Abstract

Since 2010, the same standard has been used in EU countries, which includes design load models for road bridges. However, some countries have adopted load factor values other than those recommended in the standard, which has resulted in large differences in the design load of bridges, including bridges located on trans-European network routes. In addition, the standard design load is close to that of a column of actual heavy vehicles when congestion occurs on a bridge. This chapter proposes the adoption of a uniform increased design load for bridges, guaranteeing the safe transfer of European traffic. Currently, vehicle traffic on the trans-European road network traverses thousands of bridges designed to different and outdated load standards. The future of these bridges depends on what load is adopted when checking their service load capacity. This chapter proposes adopting a service load based on the characteristics of the vehicles allowed on European roads.

Keywords: road bridges, trans-European road network, bridge design load capacity, bridge service load capacity, standard design load

1. Introduction

The smooth operation and development of the European market and Europe's economic, social and territorial cohesion depend on the functioning of the trans-European transport networks. These networks concern road, rail, maritime and inland waterway traffic. Within the transport networks, the road network occupies an important place. There are thousands of bridges and viaducts (hereafter referred to as "road bridges", "bridges" or "structures") on the main public roads in Europe that make up the trans-European road network. The usability of these bridges is determined by their load-bearing capacity. Some bridges with a load capacity sufficient to carry the current traffic will be able to continue to be used without reconstruction; some will require reconstruction (strengthening or widening), and some will be demolished, and new bridges will be built in their place. How a particular road bridge in service should be dealt with: leave it without reconstruction, reinforce it or demolish it, depends on both its design load capacity and its service load capacity. These are two load-bearing capacities that depend on each other, but are completely different.

1.1 Design load capacity and service load capacity of a bridge

The design load capacity of road bridges is the load capacity that the structure has when designed to a given load standard and for the load class specified in the standard. Until 2010, each European country had its own national load standards. So, today, vehicles on trans-European road network traffic over bridges are designed to meet different standards. As a rule, the later the standard was issued, the higher the design load of the bridge. Without reconstruction, a bridge has (and will always have) only one design load capacity – the one it was designed for, regardless of which load standard it was designed to, for which load class and over what period.

Since 1 April 2010, in Europe, only the European standards for the design of structures for civil engineering works, the so-called Eurocodes [1, 2] with their national annexes [3–6], have to be used for the design of bridges. These standards are currently used in 33 countries that are members of the European Committee for Standardization.

Road bridges have a service load capacity after commissioning. It can be assumed that the load-carrying capacity of a bridge is the load of vehicles with such a total vehicle weight that the load-carrying capacity and serviceability conditions of the structure are met. If the weight of the vehicles allowed to traffic on the structure is less than the weight of the vehicle allowed to traffic on the public road without restrictions, the bridge should be marked. The sign shall bear a number indicating the actual total weight, in tons, of the vehicle authorized to use the structure. Thus, the carrying capacity of a bridge is determined by the weight of the vehicle allowed to drive on that structure without restrictions. In Poland, the rules for determining the load capacity of road bridges were developed by the author of this chapter and recommended for use in 2004 by the General Director of National Roads and Motorways [7].

1.2 Difference between design load and service load of a bridge

Fundamental to the determination of both the design load and the service load of bridges are the load models. The variable design load for road bridges is specified in EN 1991–2 Eurocode 1 [2]. This standard gives the values of the forces and how the design load models are located on the roadway of the structure. These models represent a simulation of road traffic.

The design load should be used to determine the service load capacity of bridges in service. Eurocode 1 [2] provides some information related to service load but does not provide models for such load. In this situation, it is sometimes assumed that bridges in service should meet the design requirements and carry the design load. Since the internal forces in a bridge structure under design load are generally greater than under service load, the use of design requirements to determine the service load capacity of in-service structures is unjustified. An in-service structure designed to an outdated standard will never meet the current design requirements because it was designed for a lower load.

According to the Regulation of the European Parliament and of the Council [8], each structure during its foreseeable lifetime should have an adequate load-bearing capacity and should ensure safety in use. For structures in service, the safety of their use should be ensured by specifying the conditions of that use, that is, the indication of the service load, including the maximum weight of the vehicle allowed to traffic on the structure without restriction, and the positioning on the roadway of the vehicles allowed.

2. Design load models for bridges according to the European standard

Variable design load models for bridges are contained in the European Standard Eurocode 1 [2] (in this chapter, the term "European Standard" or "Standard", without reference to a bibliographic entry, means Eurocode 1 [2]). It should be noted that, in discussing design loads, the provisions of this standard have been used, giving them in inverted commas without any authorial commentary.

In the standard, the design load models have been adopted in such a way that their impact is that of "the actual in the year 2000 traffic in European countries" (Section 4.2.1(1)). A dynamic surplus (dynamic factor), determined "for a medium pavement quality and pneumatic vehicle suspension" (Section 4.2.1(1)), is included in the load values.

The load values are given as characteristics. According to the European Standard for the fundamentals of structural design, the characteristic value of a variable action corresponds to a nominal value, which can be taken when the statistical distribution is unknown. Thus, the values of loads as characteristic values will be accepted for analysis without additional safety factors ([1], Section 4.1.2(7)).

Four vertical force design load models are given in the standard:

- Model 1, which is the basic model and concerns the loading of the bridge with vehicles;
- Model 2 concerns the loading of the bridge with one vehicle axle;
- Model 3 is for the loading of the bridge with special standard vehicles;
- Model 4 concerns the loading of the pavement with a crowd of pedestrians.

Since, of the above four models, Model 1 is the basic loading model and is related to vehicle loading, and this model was adopted for further analysis.

2.1 Characteristics of the basic load model

Model 1 consists of a characteristic load (indicated by the subscript "k" when describing the loading) of the roadway with concentrated forces Q_{ik} and a uniformly distributed load q_{ik}, where "i" is the lane number. In this model, the concentrated forces are two axles (tandem *TS*) located in each lane. The load values are given in **Table 1**.

Lane No.	Load Model 1	
	Q_k – concentrated forces - tandem *TS*	q_k – uniformly distributed load *UDL*
	[kN]	[kN/m^2]
1	2 × 300	9.0
2	2 × 200	2.5
3	2 × 100	2.5
>3	—	2.5

Table 1.
Characteristic load values in model 1.

According to the standard, the lane width is 3.00 m, the tandem wheelbase is 1.20 m and the wheelbase per axle is 2.00 m. Only full tandem sets are to be considered for loading, and uniformly distributed loads are to be placed at the most unfavorable point of the impact surface. The order in which the lanes are positioned on the roadway should be such that the effects caused by the loads are most unfavorable (Section 4.2.4(2)). For span lengths greater than 10 m, the tandem arrangement is replaced by a single-axle concentrated load equal to the sum of the two axle loads (Section 4.3.2(6b)). A diagram of the load model, in side and top views, is shown in **Figure 1**. The load, according to Model 1, is the product of the characteristic load and the adjustment factors.

2.2 Values for the adjustment factors recommended in the European standard

In the European Standard, the adjustment factors are given for the characteristic load in Model 1:

- α_{Qi} – for concentrated forces – tandem system *TS*;

- α_{qi} – for uniformly distributed load *UDL*.

The standard recommends the following minimum adjustment factors: $\alpha_{Qi} \geq 0.8$ and $\alpha_{qi} \geq 1.0$, with $i \geq 1$ (Section 4.3.2(3)). "For road bridges, Load Models 1 and 2 (...) and taken into account with adjustment factors α and β equal to 1, are deemed to represent the most severe traffic met or expected in practice, other than that of special vehicles requiring permits to travel, on the main roust of European countries" (foreword). With this value of factors, "a heavy industrial international traffic is expected, representing a large part of the total traffic of heavy vehicles. For more

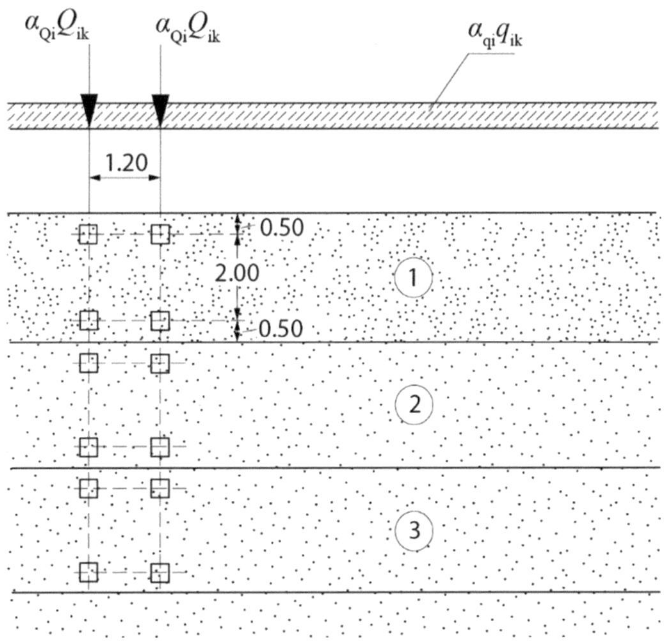

Figure 1.
Diagram of load model 1.

common traffic compositions (highways or motorways), a moderate reduction of α factors applied to tandem systems and uniformly distributed loads on Lane 1 may be applied (10 to 20%)" (Section 4.3.2(3)).

In Table 2.1 of the European Standard, the characteristic load values in Model 1 are assumed such that, with adjustment factors α equal to 1.0, the probability of them being exceeded on Europe's major roads in 50 years is 5%, indicating a "1000-year return period."

2.3 Values for adjustment factors adopted in some European countries

Nationally Determined Parameters are allowed in the European Standard. These parameters are given in the national annex to the standard (national annexes are denoted by the acronym NA). Such parameters include the adjustment factors α_{Qi} and α_{qi}, which increase or decrease the characteristic load in Model 1.

In European countries, including EU member states, the national annexes to the standard have adopted the adjustment factors α_{Qi} and α_{qi} with different values [9]. In some countries, factor values of 1.00 have been adopted, and in some countries, such as Denmark, France, Germany, Poland and the United Kingdom, among others, values other than 1.00 have been adopted.

Table 2 summarizes the adjustment factors used in some European countries for the design of bridges located on the trans-European road network.

Analyzing the values of the factors given in **Table 2**, it can be seen that:

- in Denmark, the value of the factor for the load uniformly distributed on the first lane was reduced by 33%;
- France increased the factor for the load uniformly distributed on lanes other than the first lane by 20%;
- in Germany and Poland, the factors for the load uniformly distributed on all lanes have been increased by 33% for the first lane, 140% for the second and 20% for the other lanes;

European country	Adjustment factors				
	α_{Qi}		α_{qi}		
	$i=1$	$i>1$	$i=1$	$i=2$	$i>2$
Most countries	1.00	1.00	1.00	1.00	1.00
Denmark [5]	1.00	1.00	0.67	1.00	1.00
France [6]	1.00	1.00	1.00	1.20	1.20
Germany [4]	1.00	1.00	1.33	2.40	1.20
Poland [10]	1.00	1.00	1.33	2.40	1.20
United Kingdom [3]	1.00	1.00	0.61	2.20	2.20

Designations adopted in the table:
i – lane number;
α_{Qi} – adjustment factor for concentrated forces;
α_{qi} – adjustment factor for uniformly distributed load.

Table 2.
Summary of adjustment factors for the design of bridges over the trans-European road network.

- in the United Kingdom, the factor for the load uniformly distributed on the first lane was reduced by 39% and the factor for the other lanes was increased by 120%.

The adoption of different adjustment factors by different European countries (in the appendixes to the standard) results in different design load values being adopted for bridge design in these countries. This, in turn, results in different design loads for bridges, that is, European road traffic is not provided with the same level of safety. For example, if the values of the adjustment factors given in **Table 2** are adopted, with design loads according to the same European standard, the load-carrying capacity of bridges designed in Denmark will be lower than those designed in France, and in France will be lower than those designed in Germany and Poland [9].

3. Proposal for adoption of increased equal design load by European countries

Equal loading of bridges designed on the trans-European road network can be achieved by adopting equal values for the adjustment factors in Model 1. In this respect, there should be no freedom if traffic in Europe were to be carried by structures with the same level of road safety.

The study [11] gives the history of the work on the European load standard. Work on the standard started in 1987, but traffic surveys were carried out in several European countries as early as 1977. It was these surveys that formed the basis for the adoption of design load models, including Model 1. It should be noted that almost 50 years have passed between the time the surveys were carried out and the present. During this period, traffic has changed fundamentally: vehicles have more weight and their share of traffic is greater than it was a few decades ago [12].

In the Eurocode design system, according to Table 2.1 in EN [1], the design of new bridges should take into account a minimum service life of 100 years. In this case, suitably high design loads should take into account the progression of service loads over this period and the progressive loss of service properties over time, while maintaining the assumed operation and ongoing maintenance of the structure.

Given that:

- Model 1 design load for bridges was adopted from traffic studies of 50 years ago,

- bridges designed on the basis of this load model should have a service life of at least 100 years,

European countries in their annexes to the standard should adopt adjustment factors that will result in the highest possible values of internal forces induced by the design load. Such factor values were adopted in Germany and Poland – these factor values were adopted for the analysis.

Table 3 summarizes the concentrated force (*TS*) and uniformly distributed load (*UDL*) values, with the standard factors α_{Qi} and α_{qi} having a value of 1.00 and the proposed increased factors, with different values for each lane.

In characterizing the proposed design load, it can be seen that the values of the adjustment factors for concentrated forces are as recommended in the standard – for all lanes, they have a value of α_{Qi} = 1.00, while the values of the adjustment factors for

No.	Standard design load				Proposed design load			
	α_{Qi}	*TS*	α_{qi}	*UDL*	α_{Qi}	*TS*	α_{qi}	*UDL*
	[−]	[kN]	[−]	[kN/m2]	[−]	[kN]	[−]	[kN/m2]
1	1.00	2 × 300	1.00	9.0	1.00	2 × 300	1.33	12.0
2	1.00	2 × 200	1.00	2.5	1.00	2 × 200	2.40	6.0
3	1.00	2 × 100	1.00	2.5	1.00	2 × 100	1.20	3.0
>3		—	1.00	2.5		—	1.20	3.0

Table 3.
Characteristic load values in model 1 with standard factors and proposed increased factors.

uniformly distributed load have higher values than the standard – for the following lanes, the values are respectively: $\alpha_{q1} = 1.33$, $\alpha_{q2} = 2.40$ and $\alpha_{q>2} = 1.20$.

3.1 Assumptions made for the analysis

1. It was assumed that the comparison of the standard design load with the proposed design load (Model 1 with different values of adjustment factors) is based on the comparison of internal forces – bending moments and transverse forces – in the load-bearing member of the bridge span structure.

2. Bridges with the static scheme of a simply supported beam were assumed. In spans with this static scheme, the internal forces in the load-bearing members of the structure that determine the strength of the section are usually bending moments and transverse forces.

3. Spans up to l = 200 m were assumed, as this span was determined by calibration of the basic design load – Model 1 (Section 4.1(1)).

4. Due to the analyzed spans up to 200 m, beam span cross-sections (without walkways) and the number of girders equal to: n = 2, 4, 6 and 12 were assumed.

5. It was assumed that the analyzed load-bearing element of the bridge span structure is the outermost girder, as it follows from the analysis of the transverse load distribution of the beam span and from the design practice that under standard loading, the outermost girder is usually the most overloaded load-bearing element in the span (the mutual location of the axis of this girder and the edge of the roadway determines this).

6. The usable width of the roadway has been assumed to be b = 6.00, 9.00 and 12.00 m, as, according to Table 4.1. in the standard, the width of the conventional lane is 3.00 m – that is, a two-lane, three-lane and four-lane roadway has been assumed, respectively.

7. The axial spacing of the outermost girders "*s*" has been assumed such that each girder has a roadway lane of the same width (e.g., if n = 2 and b = 6.00 m, then s = 3.00, and if n = 6 and b = 9.00 m, then s = 7.50 m).

8. In order to compare the internal forces in the extreme girder of the span structure, the forces arising in a simply supported beam loaded as individual lanes were multiplied by the ordinates of the line of influence.

9. The values of the ordinates of the lines of influence of the transverse load distribution in the structures were calculated using the *Courbon* method [13]. This is a method that assumes that there is a transverse beam with infinite stiffness at the center of the span. It is assumed that with a span-to-width ratio of the bridge span of not less than 2, the results obtained are correct. On the other hand, it should be noted that in this chapter, the loads are compared and in this case, the values of the ordinates of the transverse section are not so relevant.

10. As the horizontal forces (braking and acceleration) "should be calculated as a fraction of the total maximum vertical loads corresponding to Load Model 1" (4.4.1(2)), only the vertical loads were analyzed.

11. The loads given in the standard are expressed in kilonewtons (kN). However, due to the stipulation in Article 2 of the Council Directive [14] that a ton means the weight due to a load of one ton corresponding to 9.8 kilonewtons and the expression in this Directive in tons of the permissible mass of the vehicle per axle, in the following analysis the mass per axle is expressed in tons and denoted by the letter "*t*".

3.2 Analysis of standard design load and proposed increased load

For the analysis, the standard design load – Model 1 – was adopted with adjustment factors:

- those recommended in the standard: both the value of the factor for concentrated forces α_{Qi} and the factor for uniformly distributed load α_{qi} is 1.00;

- proposed for adoption by European countries (already adopted by Germany and Poland): the value of the factor for concentrated forces α_{Qi} is 1.00, and the value of the factor for uniformly distributed load α_{qi} is 1.33 for the first lane, 2.40 for the second lane and 1.20 for the third and other lanes.

Table 4 gives the ratio of the internal forces in the extreme girder of the span at the proposed increased design load (with adjustment factors of a higher value than

l	Width of bridge deck		
[m]	2 × 3 m = 6 m	3 × 3 m = 9 m	3 × 3 m = 12 m
50	1.19–1.20	1.21–1.24	1.23–1.24
100	1.25–1.26	1.28–1.32	1.30–1.33
150	1.27–1.28	1.32–1.36	1.34–1.37
200	1.29–1.30	1.34–1.39	1.37–1.40

Table 4.
Ratio of internal forces under loading proposed and standard.

the standard), compared to the design load with standard factors. The force values were compared for spans equal to l = 50, 100, 150 and 200 m. The force ratios apply to both bending moments and transverse forces and to spans with different numbers of beams (n = 2, 4, 6 and 12).

Figure 2 shows the values of the internal forces in the outermost girder of the bridge span, with roadway widths of b = 6, 9 and 12 m and with the number of beams n = 2. The internal forces are bending moments (as unit forces, i.e., bending moments divided by the span). In **Figure 2**, the design load with factors $\alpha_{Qi} = \alpha_{qi} = 1.00$ recommended in the European standard is denoted "EU" and the design load with factors proposed to be uniformly acceptable in European countries is denoted "PL" (this is an acronym for "proposed load").

Analyzing the data summarized in **Table 4** and **Figure 2**, it can be concluded that the ratio of loads practically does not depend:

- on the type of internal force, that is, in a beam span, the ratio of bending moments is the same as the ratio of transverse forces, induced by these loads;
- on the number of girders in the cross-section of the span, that is, the ratio of bending moments in a 2-girder span is the same as that in a 12-girder span induced by these loads.

Moreover, analyzing the data summarized in **Table 4**, it can be concluded that:

- the larger the span is, the greater is the impact of the increased value of the factors, that is, increasing the span from l = 50 to l = 200 m, with a two-lane roadway, the value of the internal force increases from 19–30%, and with a four-lane roadway from 23–40%;

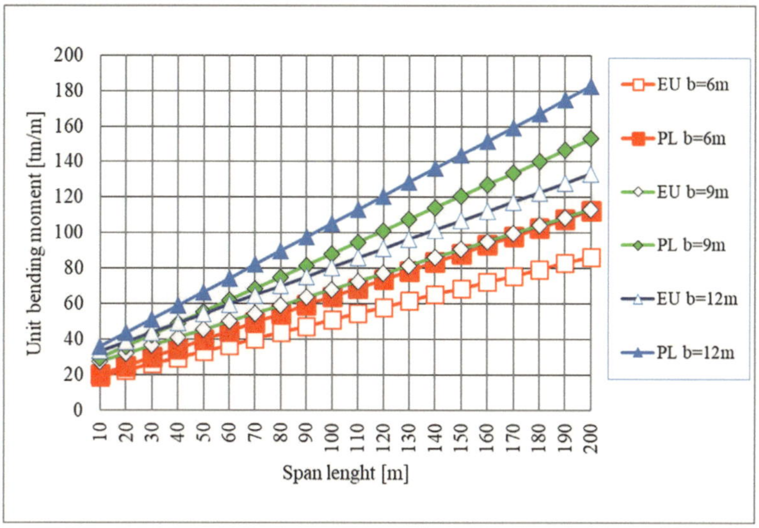

Figure 2.
Values of bending moments in the outermost girder of a bridge span with roadway widths b = 6, 9 and 12 m and number of beams n = 2, under standard and proposed design loads.

- the greater the number of lanes on the roadway, the greater the impact of the increased value of the factors, that is, for example, with a span of l = 100 m, the value of the internal force with a 2-lane roadway increases by about 25%, and with a 4-lane roadway by about 30%.

Since – under the assumptions made – the load ratio practically depends neither on the type of internal forces adopted nor on the number of girders, in further analysis, only the lateral forces in a 2-girder structure and – solely for the purpose of checking the calculations – a 12-girder bridge span structure were compared.

4. Characteristics of lorries according to the council directive and according to the European standard

The Council Directive [14] specifies the maximum weights and axle loads of vehicles allowed in national and international traffic in Europe. Vehicle weights depend on the number of axles. The maximum vehicle weight is:

- 26 t, if the vehicle has 3 axles;
- 32 t, if the vehicle has 4 axles;
- 40 t, if the vehicle has 5 axles.

The axle loads of a vehicle depend on the type of axle (powered or non-powered) and its wheelbase. The maximum axle load of a vehicle is:

- 11.5 t if the axle is driven and single;
- 19 t (2 × 9.5), if the axle is drive and double, or if one axle is drive and the other is non-drive, and the wheelbase is not less than 1.3 m;
- 10 t if the axle is non-drive and single;
- 24 t (3 × 8), if the axle is non-powered and triple, and the wheelbase is greater than 1.3 m;
- 21 t (3 × 7), if the axle is non-driven and triple, and the wheelbase is no greater than 1.3 m.

Table 5 summarizes the parameters of lorries that meet the requirements of the Council Directive [14]. The vehicles in this table are given under the numbers 2, 4, 6 and 7. In the column where the weight of the vehicle is summarized, the number of axles of the vehicle is also given with the letter "R", indicating the real vehicle.

In the European standard, the fatigue load model gives sets of "standard lorries which together produce effects equivalent to those of typical traffic on European roads" (Section 4.6.5).

Table 5 summarizes the parameters of the lorries given in Table 4.7 of the standard. The standard gives the parameters of two five-axle vehicles: an articulated vehicle, consisting of a motor vehicle and a semi-trailer, and a combination of

No.	Lorry silhouette	M	P	S	S_i	M/S
		[t]	[t]	[m]	[m]	[t/m]
1	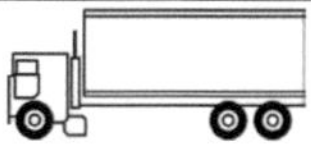	31.6 3 N	7.2 12.2 12.2	5.50	4.20 1.30	5.75
2		26 3R	7.0 9.5 9.5	4.50	3.05 1.45	5.78
3	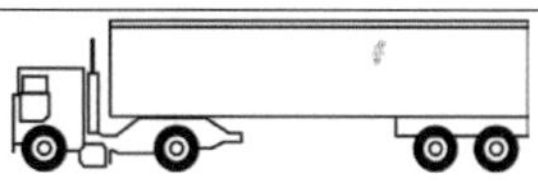	39.8 4 N	7.1 14.3 9.2 9.2	11.20	3.40 6.00 1.80	3.55
4	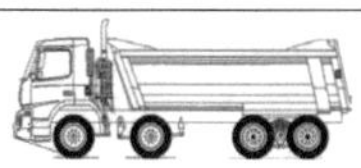	32 4R	6.6 6.6 9.4 9.4	5.55	1.70 2.50 1.35	5.77
5	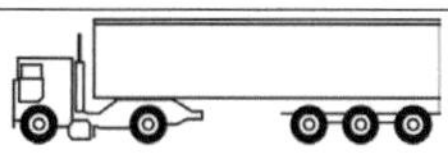	A50 5 N	7.1 15.3 9.2 9.2 9.2	11.00	3.20 5.20 1.30 1.30	4.55
6	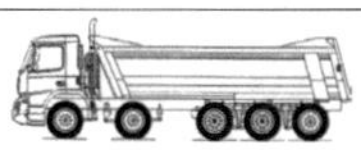	40 5R	7.0 7.0 9.0 9.0 8.0	7.10	2.00 2.40 1.35 1.35	5.63
7	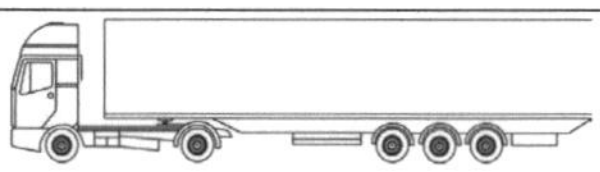	A40 5R	6.5 11.0 7.5 7.5 7.5	12.00	3.60 5.70 1.35 1.35	3.33

Designations adopted in the table:
M – vehicle weight;
N – standard vehicle;
P – axle loads of the vehicle;
R – real vehicle;
S – spacing of extreme vehicle axles;
S_i – spacing of individual vehicle axles.

Table 5.
Summary of lorry parameters actual and standard.

vehicles, consisting of a motor vehicle and a trailer. Due to the fact that an articulated vehicle has a higher weight and shorter wheelbase than a combination of vehicles, the standard articulated vehicle (marked with the letter "A") is characterized in **Table 5**. The standard vehicles in **Table 5** are given under numbers 1, 3 and 5 (these vehicles are marked with the letter "N", indicating the standard vehicle).

4.1 Comparison of vehicle weights and axle loads according to the council directive and the European standard

Table 5 summarizes the parameters of heavy-duty vehicles that meet the requirements of the Council Directive (Nos. 2, 4, 6 and 7) and the European standard (Nos. 1, 3 and 5).

Analyzing the characteristics of the vehicles given in the standard, it can be concluded that their parameters are inconsistent with those given in the Council Directive [14]:

- vehicle No. 1 – three-axle, has a mass of 31.6 t, which is 22% greater than the maximum mass of 26 t, and a double axle load (one drive) of 24.4 t (12.2 t × 2), which is 28% greater than the maximum load of 19 t, while its extreme axle spacing of 5.50 m is greater than that of some actual three-axle vehicles (e.g., vehicle No. 2 has an extreme axle spacing of 4.50 m);

- vehicle No. 3 – a four-axle vehicle, has a weight of 39.8 t, which is 24% greater than the maximum weight of 32 t, a single axle load of 14.3 t, which is 24% greater than the maximum load of 11.5 t, while the extreme axle spacing of 11.20 m is greater than actual four-axle vehicles (e.g., vehicle No. 4 has an extreme axle spacing of 5.55 m);

- vehicle No. 5 – a five-axle vehicle, has a weight of 50 t, which is 25% greater than the maximum weight of 40 t, a single axle load of 15.3 t, which is 33% greater than the maximum load of 11.5 t, and a non-driving triple axle load of 27.6 t (9.2 t × 3), which is 31% greater than the maximum load of 21 t.

Taking into account the fact that the parameters of vehicles given as standard in the standard have parameters that are incompatible with those of vehicles allowed on European roads under the Council Directive [14], real vehicles that are in traffic on European roads were adopted for further analysis.

4.2 Comparison of single lane loading with actual vehicles arranged in a column of 5.0 m intervals

In different European countries, the appendix to the European standard adopts different distances between the extreme axle of a given standard vehicle and other vehicles. For example, in the UK, 5 m was adopted [3], and in France – not less than 10 m [6]. The distance proposed in the British appendix to the standard of 5 m was adopted for further analysis.

In order to compare the loads of real vehicles of different weights, the internal forces arising in a simply supported beam loaded as if it were a 3 m wide lane were compared. Adopting the static scheme of a simply supported beam for analysis, and determining the internal forces arising when it is loaded, is in accordance with the rules given in the NATO STANAG 2021 standardization agreement [15] for determining the military load class of bridges.

Figure 3 shows the values of internal forces in the beam when one lane is loaded with real vehicles located in a column at 5 m intervals between the front axle of a vehicle and the rear axle of the next vehicle. Real vehicles weighing 26 t, 32 t, 40 t and A40t were analyzed. These are vehicles with the parameters summarized in **Table 5** under the numbers, respectively, 2, 4, 6 and 7.

Table 6 gives the ratio of the values of internal forces in the beam when loaded with real vehicles of different weights, relative to the loading of 40 t articulated vehicles (marked "A40t" in **Figure 3**). The force values were compared for spans equal to l = 50, 100, 150 and 200 m.

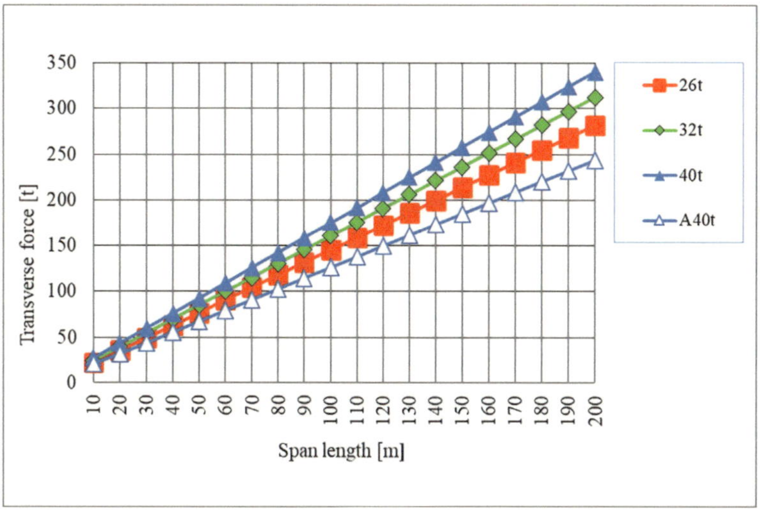

Figure 3.
Values of internal forces in the beam when one lane is loaded with real vehicles at 5 m intervals.

l	Weight of the vehicle		
[m]	26 t	32 t	40 t
50	1.14	1.27	1.38
100	1.15	1.28	1.39
150	1.16	1.28	1.40
200	1.16	1.28	1.40

Table 6.
The ratio of internal force values when loaded with real vehicles and articulated vehicles weighing 40 t.

Comparing the load on a column of real vehicles used primarily to transport construction materials (hereafter referred to as "construction vehicles"), with the load on a column of articulated vehicles weighing 40 t (3.33 t/m), it should be noted that:

- a column of vehicles weighing 26 t (5.78 t/m) induces 14–16% higher internal forces than a column of articulated vehicles;
- a column of 32 t (5.77 t/m) vehicles induces 27–28% greater internal forces than a column of articulated vehicles;
- a column of 40 t (5.63 t/m) vehicles induces 38–40% greater internal forces than a column of articulated vehicles.

The higher value of internal forces in the beam, when loaded with vehicles of lower weights (26 and 32 t), is due to the fact that the unit load of these vehicles (vehicle weight divided by wheelbase), given in **Table 5** and above in parentheses, is greater than the unit load of an articulated vehicle. In order for the mass of a vehicle to determine the value of internal forces induced by the load on that vehicle (the greater the mass, the greater the value of internal forces in the structure) in the European Council Directive [14], one of the parameters characterizing the vehicle should be the maximum unit load.

5. Comparison of the standard and proposed increased design load with the actual 40 t vehicle load

Assumptions made for the analysis.

1. The standard does not specify either the distance between the standard vehicles in a lane or the way they are seated in each lane. However, the distance between a vehicle and another load is given. It is 25 m (A.3(6)). This chapter assumes that this is the distance between the extreme axles of adjacent vehicles in traffic.

2. Lane 1 was loaded with real vehicles weighing 40 tons in two ways:

   - vehicles are located in a column, and the distance between the extreme axles of adjacent vehicles is 5 m (as in [3]). Such loading is in accordance with the standard's stipulation that Load Model 1 takes into account "traffic jam situations with a high percentage of heavy lorries" (Section 4.3.2(1b)). Due to the short distance between vehicles, the load was not increased by a dynamic factor. The load is indicated by the symbol "40 t/5 m" or "A40t/5 m";

   - vehicles traffic in a column and the distance between the extreme axles of adjacent vehicles is 25 m. This distance between the extreme axle and another load is recommended by the standard (Section A.3(6)). Such a load will be treated as occurring in traffic "with a high percentage of heavy lorries". In this case, the load was increased by a dynamic factor. The load is indicated by the symbol "40 t/25 m" or "A40t/25 m".

3. Lanes other than lane No. 1 were loaded with real vehicles of 40 t moving in a column, and the distance between the extreme axles of adjacent vehicles is 50 m. Such a load will be treated as occurring in traffic on lanes loaded with lorries, but to a lesser extent than lane No. 1. In this case, the load was increased by a dynamic factor.

4. The dynamic factor recommended in the standard (p. A.3(5)) with a value calculated according to the formula was adopted for vehicle models that move at a speed equal to 70 km/h:

$$f = 1,40 - \frac{L}{500} \tag{1}$$

in which the following designations are adopted:

f – dynamic surplus (dynamic factor),
L – influence length (in this chapter, it is the span) [m].

5. Taking into account the conclusions of the previous analysis, the transverse forces were compared in a 2-girder structure and – only to check the calculations – a 12-girder span structure.

Table 7 gives the ratio of the internal forces on the roadway with widths b = 6, 9 and 12 m, with a standard load (marked "EU") and the internal forces induced by the loading of a column of articulated vehicles weighing 40 t (marked "A40") or a column of construction vehicles weighing 40 t (marked "40"):

- on the first lane located every 5 m or every 25 m;
- on the remaining lanes located every 50 m.

The ratio of internal forces from these loads is given for spans equal to l = 50, 100, 150 and 200 m.

It should be noted that **Figure 4** shows the values of internal forces in the extreme girder of the bridge span, with a roadway width of 6 m. Obviously, these values are higher at roadway widths of 9 and 12 m, but the ratio of force values at the proposed and standard loads is similar, regardless of the roadway width (you can compare the values in the last column of **Table 7**). For this reason, the values of forces at a different span width than 6 m are not shown.

l	**UE/A40/25 m**	**UE/A40/5 m**	**UE/40/25 m**	**UE/40/5 m**	**PL/UE**
[m]	*b* = 6 m				
50	2.40	1.97	2.06	1.43	1.20
100	2.44	1.62	2.16	1.16	1.26
150	2.62	1.49	2.27	1.07	1.28
200	2.82	1.42	2.46	1.02	1.30
	b = 9 m				
50	2.05	1.76	1.81	1.36	1.24
100	2.09	1.50	1.87	1.13	1.32
150	2.23	1.41	1.96	1.04	1.36
200	2.40	1.36	2.12	1.00	1.39
	b = 12 m				
50	1.86	1.64	1.66	1.31	1.24
100	1.91	1.45	1.73	1.11	1.33
150	2.05	1.38	1.82	1.04	1.37
200	2.22	1.35	1.97	1.01	1.40

Table 7.
The ratio of the values of internal forces under load standard and a column of vehicles weighing 40 t.

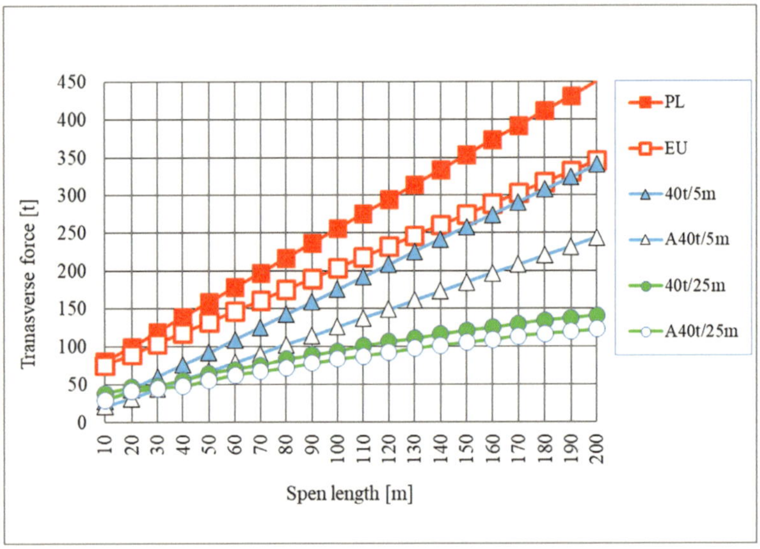

Figure 4.
Values of internal forces in the outermost girder of a bridge span with a roadway width of b = 6 m when loaded with actual 40 t vehicles and with the standard and proposed design loads.

Comparing the values of internal forces shown in **Figure 4** and the ratio of internal forces under the standard load and 40 t vehicles, it should be noted that the load according to the European standard induces internal forces:

- 35–97% greater than the forces induced by a column of articulated vehicles weighing 40 t and located every 5 m;

- by 0–43% greater than the forces induced by a column of construction vehicles weighing 40 t and located every 5 m.

It should be noted that the larger the span is, the difference in internal forces from the standard load and vehicles weighing 40 t (at 5 m intervals), is smaller. For example, with a roadway width of 6 m and a span of 50 m, the internal forces from the standard load are higher by 43%, and with a span of 200 m only by 2%. However, with a roadway width of 9 m and a span equal to 50 m, the standard forces are 36% greater, and with a span equal to 200 m, equal to a vehicle load of 40 t. In addition, with a span of up to 150 m, the greater the width of the roadway, the smaller the difference in internal forces.

Annex B of the standard, which deals with fatigue life assessment of road bridges, states "a factor 1.4 for the load levels are recommended". This value can be taken as a minimum when evaluating the value of internal forces induced by the design load relative to internal forces induced by the service load. In addition, since "the characteristic loads associated with special vehicles should be taken as nominal values" (Section A.2 (3)), design and vehicle loads should be compared without using design factors.

In summary, it can be said that the load, according to the European standard, is close to that of real 40 t construction vehicles located in one lane every 5 m, and in the other lanes every 50 m (**Table 7** shows in gray the ratio of forces less than 1.4).

Moreover, the ratio of these forces is less than 1.4, even for 40 t articulated vehicles. Keeping in mind the expected service life of bridges according to the European standard [1], certainly, such a design load, which can be replaced by 40 t real vehicle loads, cannot guarantee the use of structures for 100 years.

Table 7 also gives the ratio of internal forces on roadways with widths b = 6, 9 and 12 m, under the proposed load (with increased adjustment factors for uniformly distributed load, marked "PL") and the standard load (marked "EU"); these values are also given in **Table 4**. The proposed standard load induces at least 33% higher internal forces (with a 6 m roadway and a span of 200 m) compared to the load of a 40 t construction vehicle column located at 5 m intervals.

6. Proposal for adoption of equal service load by European countries

Thousands of bridges designed according to already outdated standards have been built in the tracts of the trans-European road network. Practically every European country operates bridges designed according to several load normatives, with different ones used in different countries. Eurocode 1 [2] is the first normative (a fact that needs improvement) that is used for bridge design in European countries that are European Committee for Standardization members. The treatment of structures in service depends on the load for which they were designed and, perhaps before all, on the service load to be assumed when checking its service load capacity. The service load should be expressed in terms of the maximum weight of the vehicle permitted to travel over the structure without restriction and how the permitted vehicles are located on the roadway.

The establishment of a uniform service load on the trans-European road network is necessary because of the need to create a uniform level of traffic safety just as there are unequivocally defined maximum parameters of vehicles allowed on European roads, so the service load scheme of bridges located on the routes of the trans-European road network should be unequivocally defined.

Verification of the bridge's service load capacity should be carried out using a service load. In this load, the standard vehicle is a five-axle articulated vehicle (consisting of a tractor and semi-trailer) with a weight of 40 tons and a wheelbase of 12.0 meters. These are the parameters of the vehicles most commonly found on the road. The article [16] gives the differences in traffic with lorries in Bavaria in 1984 and 2005. A German study shows that among lorries, articulated vehicles accounted for the largest number – 50% - in 2005, and their growth over the period was 2.5 times. It is currently difficult to give the percentage of articulated vehicles among all lorries on the road, since lorries include all vehicles weighing more than 3.5 tons. It would be necessary to amend the provisions of the Council Directive [...], which recognizes that a heavy vehicle is a vehicle with a permissible total weight of more than 3.5 t, which is, for example, 4 t, and the permissible total weight of a two-axle vehicle is 18 t.

For bridges with a two-lane roadway, the following standard vehicle load should be applied:

- on the first lane – vehicles located every 5 m (without dynamic factor);
- on the second lane – vehicles located every 25 m (with dynamic factor).

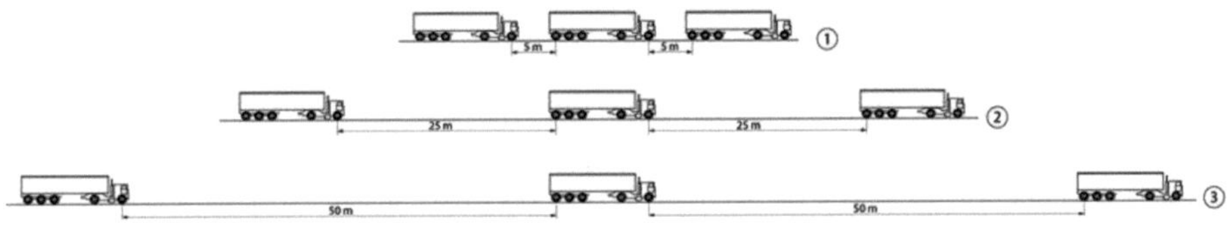

Figure 5.
Service load diagram of bridges with standard vehicles.

For bridges with a roadway of three or more lanes, the following standard vehicle load should be used:

- on the first and second lanes – as above;
- on the third and subsequent lanes – vehicles located every 50 m (with dynamic factor).

The load on the third and subsequent lanes is derived from the submission that, regardless of the traffic situation in other lanes, a standard vehicle should be able to pass over a given lane. The service load diagram for bridges with standard vehicles is shown in **Figure 5**.

Vehicles should be located in the axis of the real lane. Only a full set of vehicles should be considered for loads, and loads should be positioned at the unfavorable point of the influence surface. The order of lane positioning on the roadway should be such that the effects induced by the loads are the most unfavorable.

When checking the service load capacity, the road infrastructure manager may exchange standard vehicles for vehicles with higher unit loads in the service load scheme, but the distance between vehicles in the lanes should not be changed. In addition, it is permissible to load each lane with one standard vehicle with the parameters specified in **Table 5**, No. 7, and replace the remaining vehicles with their unit load – in the case of a standard vehicle equal to 3.33 t/m (with a distance between vehicles of 25 m, the error is no more than 2%).

Checking the load-carrying capacity is not a form of structure design. When designing a given structural element (e.g., the outermost girder), the load of individual lanes is set at the most unfavorable point of the area of influence for this element (the area of influence for another element, in a multi-girder structure, will be different). On the other hand, when checking the service load capacity, the entire bridge structure, all its structural elements, carry the load that will be placed on the roadway (for example, for a two-lane roadway, one lane is usually loaded separately and two lanes together). Thus, the design of bridges should not be confused with checking their load-carrying capacity. In addition, it should be remembered that the proposed service load will primarily apply to bridges designed according to outdated load standards.

7. Conclusions

The European standard Eurocode 1 [2] includes variable design load models for road bridges. The basic load model is Model 1. The standard allows the use of parameters defined at the national level. Such parameters are the adjustment factors for the load in Model 1. The adoption by different European countries of different values for the adjustment factors in Model 1 makes it practically impossible to design bridges in Europe that would carry European traffic with the same level of safety. It is necessary to agree at the European level on the values of adjustment factors, that is, the same design load for bridges.

Equal loading for bridges designed along the trans-European road network can be achieved by adopting equal values for the adjustment factors in Model 1.

Taking into account that:

- the standard adopts design load models appropriate for "the actual traffic in the year 2000 in European countries" (Section 4.2.1(1)), and a quarter of a century has passed since then and traffic has changed fundamentally;
- the load according to the European standard is close to that of a column of real construction vehicles of 40 t;
- such design load, which can be replaced by the load of real vehicles of 40 t, does not guarantee the use of structures for 100 years.

adaptation factors with higher values than those recommended in the standard should be adopted in all European countries. Such values of factors have already been adopted in Germany and Poland (when applying these factors, for example, increasing the span from l = 50 to l = 200 m, with a 2-lane roadway, the value of internal force increases from 19–30%).

Thousands of bridges have been built on the trans-European road network, designed according to different and already outdated load norms. The handling of these in-service bridges depends on which service load is adopted when checking their service load capacity.

Determining a uniform service load on the tracts of the trans-European road network is necessary because of the need to create a uniform level of traffic safety. In the service load, the standard vehicle is an articulated vehicle with a weight of 40 t and a wheelbase of 12.0 m (these are the parameters of the vehicles most commonly found in road traffic).

For bridges with a roadway with any number of lanes, the following standard vehicle load should be used:

- on the first lane – vehicles located every 5 m;
- on the second lane – vehicles located every 25 m;
- on the third and other lanes – vehicles located every 50 m.

Equally safe and durable bridges over the trans-European transport network are structures designed for the same design load and then loaded with the same service load. The Polcevera viaduct in Genoa, Italy, collapsed when 3 lorries were caught in the support abutment [17]. On this viaduct, the same problem that is currently occurring throughout Europe – structures of excessive dimensions are being built – occurred in an extreme form. Bridges currently being built generally have large spans and the high proportion of Heavy Goods Vehicles in traffic will create large, previously unheard of service loads. One easy way to maintain the correct relationship between design and service loads is to build bridges with as small a span as possible.

The Council Directive [14] gives the maximum weights and axle weights of vehicles permitted in national and international traffic in Europe. However, taking into account the real parameters of vehicles, it turned out that vehicles with lower weights can cause greater internal forces than vehicles with higher weights (vehicles with a weight of 26 or 32 t cause greater internal forces than vehicles with a weight of 40 t). This is due to the fact that the unit load of vehicles with lower total weight (weight per wheelbase) is higher.

Such a situation is ambiguous for bridge management administrations, because until now, if a bridge could not safely carry the load of vehicles with the maximum weight allowed on public roads, traffic of vehicles with a lower total weight was usually allowed over it. And this may not ensure less strain on the structure.

Another vehicle-related parameter, the vehicle unit load, should be included in the Council Directive [14]. The maximum unit load of a vehicle should be established in Europe, similar to the maximum weight or maximum axle load. This is necessary in the context of the amendment under discussion in Europe, allowing vehicles up to 60 t in weight, up to 25.25 m in length, to operate on selected European routes. In addition, the Council Directive [14] should place other limits on which a vehicle can be considered a heavy vehicle. The existing definition of a heavy vehicle as a vehicle weighing more than 3.5 t is technically unreasonable if a two-axle vehicle can weigh 18 t. The weight of a heavy vehicle would have to be linked to the permissible axle load, for example.

Author details

Janusz Rymsza
Road and Bridge Research Institute, Warsaw, Poland

*Address all correspondence to: jrymsza@ibdim.edu.pl

References

[1] EN 1990: 2004. Eurocode: Basis of Structural Design. Brussels: European Committee for Standardisation; 2004

[2] EN 1991-2: 2003. Eurocode 1: Actions on Structures, Part 2: Traffic Loads on Bridges. Brussels: European Committee for Standardisation; 2003

[3] BS EN 1991-2/NA: 2008. UK National Annex to Eurocode 1: Actions on Structures–Part 2: Traffic Loads on Bridges. London: BSI; 2008

[4] DIN EN 1991-2/NA: 2011. Nationaler Anhang – national festgelegte Parameter–Eurocode 1: Einwirkungen auf Tragwerke–Teil 2: Verkehrslasten auf Brücken. Berlin: BEUTH VERLAG; 2011

[5] DK EN 1991-2/NA: 2009. Nationalt Anneks til. Eurocode 1: Last på bærende konstruktioner – Del 2: Trafiklast på broer. Kopenhaga: Vejdirektoratet; 2010

[6] NF EN 1991-2/NA: 2008. Annexe Nationale. Eurocode 1 – Actions sur les structures – Partie 2: Actions sur les ponts, dues au trafic. Paris: AFNOR; 2008

[7] Order No. 17 of the General Director of National Roads and Highways, On the Implementation of the Instruction for Determining the Load Carrying Capacity of Road Bridges. Warsaw: GDDKiA; 2004

[8] Regulation (EU) No 305/2011. European Parliament and of the Council of 9 March 2011 Laying Down Harmonised Conditions for the Marketing of Construction Products and Repealing. Council Directive 89/106/EEC (OJ L 88, 4.4.2011, p. 5). Luxembourg: Publications Office of the European Union; 2011

[9] Rymsza J. Proposal to change the design load in the Eurokode 1 based on loads from vehicles with a mass of 60 tonnes. Transportation Research Procedia. 2016;**14**:4020-4029

[10] Decree of the Minister of Infrastructure dated June 24, 2022 on technical and construction regulations for public roads (Journal of Laws of 2022, item 1518)

[11] Calgaro J-A, Tschumi M, Gulvanessian H. Designers' Guide to Eurocode 1: Action on Bridges. London: Thomas Telford; 2010

[12] Moving Freight with Better Tracks. International Transport Forum. Research Report. Paris: OECD; 2011

[13] Courbon J. Calcul des ponts à poutres multiples solidarisées par des entretoises. Annalies des Ponts et Chaussées, mmoires et documentes relative à l'art des constructions au service de l'ingénieur. 1940;**17**:293-322

[14] Council Directive 96/53/EC of July 25, 1996, laying down for certain road vehicles circulating within the territory of the Community the maximum authorized dimensions in national and international traffic and the maximum authorized loads in international traffic (OJ L 235, 17.9.1996, p. 59, as amended)

[15] STANAG 2021. Military Load Classification of Bridges, Ferries, Rafts and Vehicles. 7th ed. North Atlantic Treaty Organization. Brussels: NATO Standardization Agency; 2011

[16] Freundt U, Böning S, Kaschner R. Straßenbrücken zwischen aktuellem und zukünftigem Verkehr– Straßenverkehrslasten nach DIN EN 1991-2/NA (road bridges between actual and future heavy load traffic – Road traffic loads according to DIN EN 1991-2/NA). Beton-und Stahlbetonbau. 2011;**11**:736-746

[17] Rymsza J. Causes of the collapse of the Polcevera viaduct In Genoa, Italy. Applied Sciences. 2021;**11**:8098

Section 2

Modeling, Construction, and Materials

Chapter 4

VIASTRATA®: The New Frontiers of BIM for the Digitalisation and Management of Infrastructures

Salvatore Antonio Biancardo, Mattia Intignano, Francesco De Paola and Gianluca Dell'Acqua

Abstract

Building information modelling for infrastructure (I-BIM) is used for creating and managing data during the design, construction, and operations process of roads, railways, and airports. I-BIM integrates multi-disciplinary data to create detailed digital representations that are managed in an open cloud platform for real-time collaboration. The newly founded university spin-off VIASTRATA aims to develop digital information management methods and tools for the design and construction of roads, railways, airports and hydraulic works. Its scope also includes interventions on existing constructions and ultimately aims to the creation and products with high technological value and innovative BIM services. The chapter covers structured and innovative methodologies for designing, modelling and managing transport and hydraulic infrastructures leveraging BIM. To this end, the operation of an algorithm developed in visual programming language that realises the parametric model of a road dynamically from data contained in a spreadsheet is described.

Keywords: I-BIM, computational design, asset management, interoperability, VIASTRATA spin-off

1. Introduction

In the broad context of smart infrastructures, there are several technologies driving the digital revolution in the architecture engineering construction operation (AECO) field, such as building information modelling (BIM) and geographic information system (GIS). BIM for transport infrastructure is based on the realisation of 3D geometric models enriched with additional dimensions of information: 4D-time (scheduling), 5D-cost (economic project management), 6D-sustainability (environmental focus) and 7D-FM (facility management). Three further dimensions are now being discussed: 8D safety (on site and in operation), 9D lean construction (resource optimisation) and 10D industrialisation (process staging and optimisation) [1].

GIS is a cartography and spatial planning tool that allows different types of analysis depending on the data contained in intelligent geographical maps in SHP (Shapefile) format.

GIS and BIM can be integrated in a methodology that can be defined as geo-BIM, based on the use of geometric/geographic survey technologies, such as light detection and ranging or laser imaging detection and ranging (LIDAR), which records the position of millions of points in few seconds and 360° in space using laser beams, and unmanned aerial vehicle (UAV) drones that can carry both LIDAR and ultra-high-definition cameras whose photographic data can be processed into point clouds using the technique of photogrammetry. Geo-BIM applications are very popular for several scopes, including data survey for new infrastructure planning, as well as reverse-engineering workflows, heritage and monitoring [2–8].

Tools, such as GIS and BIM, open to many possibilities in terms of efficiency production enhancement, under every aspect. In fact, these digital tools are proven to improve planning, design and building operations coordination, leading to resource optimisation and saving time and money [9].

BIM is a methodology that positively affects all phases composing the lifecycle of the asset. To do so, it is very important that the information flow is continuous and consistent within planning, design, construction and facility management. For this reason, it is very important to define a common data format being a standard for interoperability. That is IFC, standing for industry foundation classes, encoding the description of knowledge, data and shapes, of a building information model [10].

At the core of BIM, there is the concept of collaboration enabled by a central intelligent digital model shared on a cloud platform with all the designers, professionals and other stakeholders involved in the project.

This approach enables to explore and validate different design ideas and what-if scenarios: for instance, in 2021, the implementation of cost-benefit analysis has been integrated into a BIM workflow to prove the visual impact on the design alternatives of a high-speed rail in southern Italy [11].

Another core principle of BIM is that the model is based on parameters. By changing the value assumed by a certain parameter, it is possible to modify a geometric feature of the model such as the width of a road lane, the pitch between the piers of a bridge, the distance between two successive lampposts, etc.

Furthermore, another enormous advantage of modelling in BIM is the possibility of correlating all the elements of the model in such a way that by changing one of them, the others adapt [12].

The BIM model, therefore, is not only a geometric representation of an infrastructure asset, but is above all an intelligent object, capable of adapting based on rules and parameters, of complying with regulatory constraints, of offering design alternatives by constituting a useful decision-support tool, of providing precise data on quantities and materials, providing a real accounting contribution, and much more [13].

Many academic publications focus on how the use of BIM can positively affect road facilities and infrastructures for their entire service life.

Researchers have been concentrating their efforts in the past few years on evaluating the advantages of employing digital tools and procedures to support road infrastructures and transport facilities throughout their whole life from performance management and maintenance to strategic planning, design and construction [13–25].

Facility management is a natural vocation for BIM as a data management tool to maximise the results of companies under budget and time constraints [26].

A digital information model can also provide a basis for complex studies and simulations in a variety of scientific fields: interaction of infrastructure with winds, noise exposure, lighting engineering, energy, environmental and social impacts and structural analyses of modelled elements [27–32].

In addition, semi-automated workflows based on the use of algorithms that are not written textually but through visual programming languages (VPL) are in full development in these years [33].

This very direction offers very interesting prospects in terms of methodology development. In fact, the adoption of BIM in business practices is an expensive and time-consuming step, considering the need to update staff, purchase materials, software, etc. The economic impact of this methodology, already non-negligible, can further emerge by benefiting from process automation techniques. In this way, BIM could be more user-friendly and even more effective, in terms of speed and accuracy.

This study provides an overview of a generic BIM workflow, enriched by the algorithms proposed by VIASTRATA, an academic spin-off affiliated with the University of Naples Federico II.

2. BIM workflow theoretical background

In recent years, several possible workflows have emerged to produce information models of linear infrastructures such as roads, railways, airport runways and hydraulic pipelines. In the following, the framework of the methodological approach to be used for the effective application of BIM to the design of linear works will be described.

A BIM workflow is based on the use of a multiplicity of software. Firstly, one must be able to process the terrain survey data. There is a wide range of software dedicated to the processing of topographic data. Next, the work shifts to what is called BIM-authoring software. In this phase, the infrastructure is modelled in its main components: planimetric axis, vertical profile, cross section, 3D solid, and information. BIM-authoring software exports the final file in various formats, also in IFC, the most important standard for BIM interoperability. There are several software packages, called BIM viewers, that allow the exploration of the digital information model but not its editing. They are, therefore, very lightweight software, mainly used for verification.

A core concept of BIM is transparency and sharing the project with all participants. There are several cloud-based platforms that allow the creation and management of a Common Data Environment (CDE), that is, an environment in which several sectoral models (architectural, structural, plant, etc.) are associated into the so-called 'federated' model. In an ideal workflow, sharing in the cloud of the project should be transversal to all work phases.

To sum it up, the work phases are:

1. Preliminary activities as analysis of the state of the sites and parameters set definition;

2. Digital terrain model (DTM);

3. Planimetric layout;

4. Vertical profile;

5. Parametric section;

6. 3D solid model;

7. Information management.

2.1 Digital terrain model

The surface that covers the terrain, without any object on it, whether natural or artificial, is called digital terrain model. The DTM is fundamental because it constitutes the starting point for the design of the horizontal and vertical components that once coordinated constitute the three-dimensional axis of the infrastructure.

It serves as the foundational framework for computational aspects, concerning elevation or slope computations, encompassing profiles, cross sections, grading and volume calculations. The DTM process entails the construction of a data structure accessible by software to promptly retrieve elevations or slopes, depicting extant or proposed conditions.

While alternative data forms can generate surfaces, three principal data types contribute to DTM construction: point data, breakline data and contour data. Point data for DTM comprise isolated X, Y and Z coordinates devoid of interconnecting features, typically representing spot elevations in contour drawings or the mass points themselves in mass points and breaklines drawings. Notably, point data necessitate an elevation or Z component amenable to processing for elevation model construction.

Breakline data delineate linear edges of site features where conspicuous grade changes occur. Effectively employed, breaklines induce contour deflections, exemplified by pavement edges, shoulders, slope tops, wall tops and water features.

Contour data constitute interconnected strings of point data within intricate objects, represented in computer-aided design (CAD) as polylines. These polylines must possess accurate Z values, either as a constant 2D polyline or a variable 3D polyline.

Most applications in civil engineering and surveying amalgamate diverse data types to formulate a comprehensive terrain model. Triangular irregular networks (TINs) emerge as a representation of a continuous surface comprising triangular facets, primarily employed as a discrete global grid for primary elevation modelling.

TINs can be generated using three categories of vector information: altitude measurements (mass points), surface continuity breaklines and surface continuity break polygons (polygon surfaces). These points encompass X, Y coordinates and Z values, serving to establish connections with the two nearest points for triangle creation. The triangulation process is underpinned by the Delaunay algorithm, ensuring that no points reside within the circumferential boundary of a triangle.

2.2 Horizontal alignment

Based on a georeferenced DTM, it is possible to start modelling the infrastructure. The planimetric layout containing information on straight roads, fixed curves and variable curves forms the fundamental basis for joining two points on the territory. Since the DTM has elevations, it will be possible to associate a vertical profile with the alignment and finish the 3D modelling with the extrusion of the cross section along the coordinate axis.

The creation of alignments encompasses various methodologies, including the generation from polylines, pipe networks and LandXML data. Alignments are fashioned through fixed, floating and free elements. Fixed elements, though seemingly static, can adapt based on dependencies on other elements, thereby maintaining a degree of flexibility. Floating elements possess one indeterminate attribute, while free elements lack constraints and derive definition from adjoining elements. The choice of element type is contingent upon design context and available data such as through points, lengths or radii. Criteria-based design features can be employed during alignment creation to ensure adherence to local standards and facilitate the identification and reporting of any violations.

Thus, alignment can be created from other objects such as polylines, reference, pipe network, LandXML file, dgn, or directly as an alignment object with the drawing tools of the chosen software. You can proceed either by initially imposing tangents and only later adding curves as an alignment edit operation, or you can proceed by drawing tangents and curves sequentially. A core feature of BIM is that geometric models are parametric, thus driven by the values that parameters take. Therefore, it is possible to edit the elements that constitute the alignment by numerically modifying the parameters, or one can use a point-and-click approach by using grip points with immediate visual feedback of the geometry. The last operations are related to the addition of labels, for example, for progressives, the dimension of significant points, radii of curves, the length of straight lines.

2.3 Vertical profile

Profiles facilitate the examination of elevation changes along a horizontal alignment, offering insights into the topographical variations.

In much the same way of horizontal alignments, the vertical profile also consists of tangents and curves, of fixed, floating and free elements. In the same graph, it will be possible to compare a multiplicity of profiles, usually the terrain profile and the design profile (**Figure 1**). Beyond the central profile, offset profiles can be established for specific features, such as waterways or ditch banks. Profile views enable the overlay of alternative horizontal alignment profiles within the same region. It is imperative for the horizontal and vertical alignments to precisely match in length for the accurate creation of the corridor.

2.4 Parametric cross section

3D modelling of linear infrastructure is achieved by extruding a plane figure along an axis. It happens in the same way for roads, bridges, railways, water pipes, sewers, etc. The modelling of the cross section takes place on a plane whose intersection with the extrusion axis is the origin of the cross-sectional diagram. One or more baselines are often associated with this point. Each software is provided with a library of objects that the user can select and drag into the diagram to compose the cross section.

Although these default libraries are well-stocked, and there are libraries dedicated to modelling objects according to local standards; in a normal workflow, there will almost always be a need to customise the cross sections by creating some component from scratch (especially in reverse-engineering operations, where the built heritage is often not very standardised). Almost, all BIM software provides the functionality to model the objects to be assembled in the cross section.

The objects constituting the individual cross-sectional components are parametric, which means that the user can modify their width, thickness, radii, distances between points, slopes, etc.

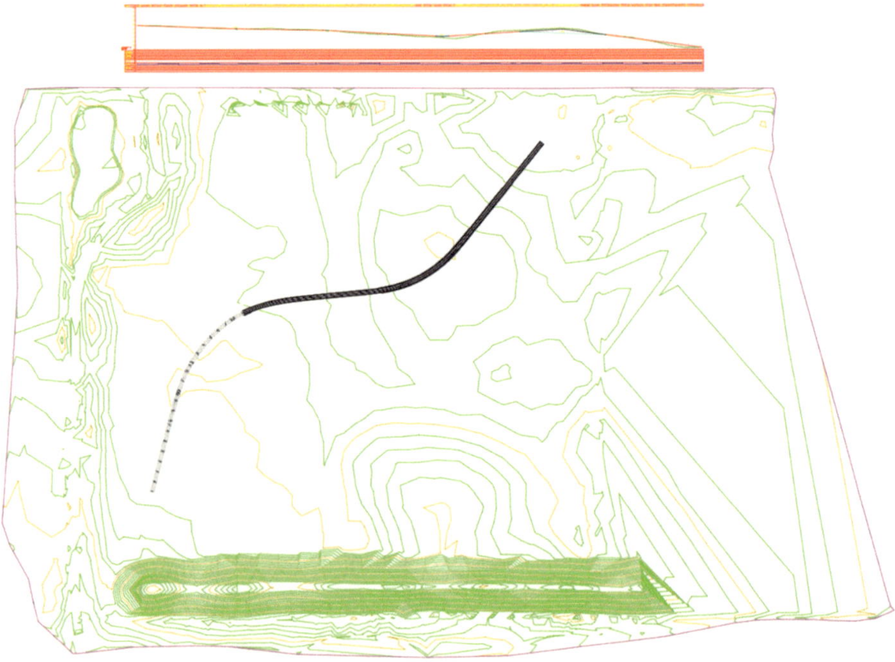

Figure 1.
DTM, road alignment and its vertical profile.

In addition, certain parameters allow to create more advanced objects. For example, targets can be associated so that certain points of the cross section can follow the course of other objects in the model, such as the DTM or breaklines.

It is also possible to programme the objects in a way that based on specific rules, they will apply to the cross section when specified conditions at a given station are met.

2.5 Corridor solid 3D geometry

The corridor is the 3D geometry representing the space occupied by the planned linear infrastructure.

At the same time as creating the corridor, it is possible to proceed with the calculation of superelevation, verification of regulations, calculation of excavation and backfill volumes, creation of print layouts and data tables, etc.

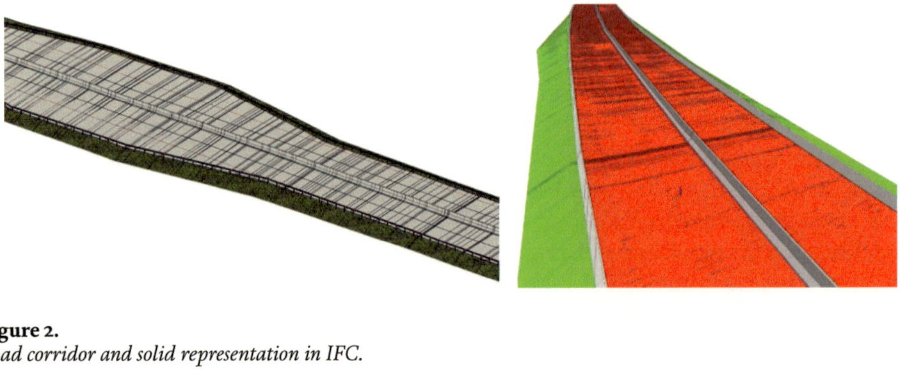

Figure 2.
Road corridor and solid representation in IFC.

Many software packages also provide water runoff analysis tools and integrate functionality for soil analysis.

The result of the workflow is, on the one hand, the production of the project drawings ready for printing, and on the other hand, the model in IFC format ready to be shared with all project stakeholders (**Figure 2**).

3. VIASTRATA university spin-off

Research on digital tools at the service of the innovative transformation of transport infrastructure engineering is in full swing. Technology transfer from the academic to the industrial and professional environment is a very important mission and a driver of scientific and economic development.

The VIASTRATA academic spin-off (**Figure 3**), affiliated with the University of Naples Federico II, has the mission to produce high-tech products and BIM services for digital construction information management and the design and construction of roads, railways, airports and hydraulic works, for new construction and for interventions on existing buildings. VIASTRATA aims to be a leader in technological innovation, introducing digital solutions that optimise the efficiency, sustainability and quality of construction worldwide [34].

The co-founders are:

Gianluca Dell'Acqua, Full Professor in Roads, Railways, and Airports and VIASTRATA Product Innovation Manager;

Salvatore Antonio Biancardo, Assistant Professor in Roads, Railways, and Airports and VIASTRATA Research and Development Manager for Transportation;

Mattia Intignano, PhD Candidate and VIASTRATA Information Technology Manager;

Francesco De Paola, Associate Professor in Hydraulics and VIASTRATA Research and Development Manager for Hydraulics;

Gerardo Attianese, Accountant and VIASTRATA Chief Executive Officer.

4. Road-BIM automatic modelling tool

The workflow described in the previous paragraphs is a proven state-of-the-art workflow to produce information models for linear infrastructures. An experienced user, depending on the complexity of the project, can still take days to complete all tasks. The proposal put forward by VIASTRATA is to speed up the entire process by using algorithms that automate individual tasks by executing them in succession and within seconds.

This is made possible by algorithms developed in a visual programming language environment (VPLE). The leading commercial software in the BIM industry provides

Figure 3.
VIASTRATA® logo.

platforms through which the average user can implement their own algorithms. The advantage of using the VPL is that no programming experience is needed to develop the algorithms. Obviously, the more complex the algorithm, the more time the developer will have to devote to it, but the more general and robust the algorithm, the wider the range of operations it can automate, and the greater the savings in time, and therefore money.

The example shown in this study is based on the use of Dynamo, an extension of Civil 3D, from Autodesk (**Figure 4**).

Figure 4 shows the diagram behind the algorithm developed by VIASTRATA for modelling a road in all its parts from the data contained in a spreadsheet. The algorithm extracts all the necessary data from an excel file, read in either CSV or XLS format, and reorganises it into editable lists.

The spreadsheet is organised in such a way that it contains data on topography, alignment, vertical profile and parametric cross-sectional geometry. These are necessary for the completion of the geometric modelling. Further, data can be entered for semantic enrichment of the model such as those relating to maintenance planning, quality, characteristics and origin, of materials, costs, structural data, laboratory test results.

The topography is summarised in a cell that provides the reference system and in a list of coordinate points with latitude, longitude and elevation, provided in separate columns.

The algorithm can read the code relating to the geographical reference system and assign it to the Civil 3D project sheet. It then draws all points according to the assigned coordinates and finally joins them by triangulation according to Delaunay.

The triangulated irregular network (TIN) surface resulting from these operations is the DTM of the BIM model. Thus, sequentially, the algorithm reads the coordinates of the points that mark the path of the alignment and the vertical profile.

Given a sequence of georeferenced points, it is possible to create a 3D polyline, resulting in a tangent break. At this point, fillet curves are defined between the tangents by specifying radii and/or length. Secondly, spirals can also be added if provided. The approach changes if the reference points are not the vertices of the project plan elements, that is, the tangents, but are surveyed points belonging to an existing road. In this second case, it is a matter of reverse engineering and making an as-built model. In this situation, if the design data are not available, one can proceed by interpolation of the points using a Non-Uniform Rational B-Spline (NURBS) curve. NURBS curves can either pass through the points entered by the user, which are called end points and are, therefore, referred to as NURBS EP or interpolate the distances between points, which are called control vertices, therefore approach the points without passing through them.

The algorithm presented can perform both procedures, depending on the input given by the user.

Once the alignment and vertical profile are defined, the algorithm creates the 'corridor' element and derives its founding elements such as baselines and regions, that is, parts of the corridor that are defined by precise chainage intervals and characterised by having the same parametric cross section, the same baselines, and generally the same representation rules.

At this point, the geometry data are read and used to create a proxy representation of the cross section within the Dynamo viewer. This geometry is then vectorised in C3D and extruded along the corridor baselines with the eventual targets assigned. This produces the 3D geometry of the road in all its components. From this geometry, the algorithm

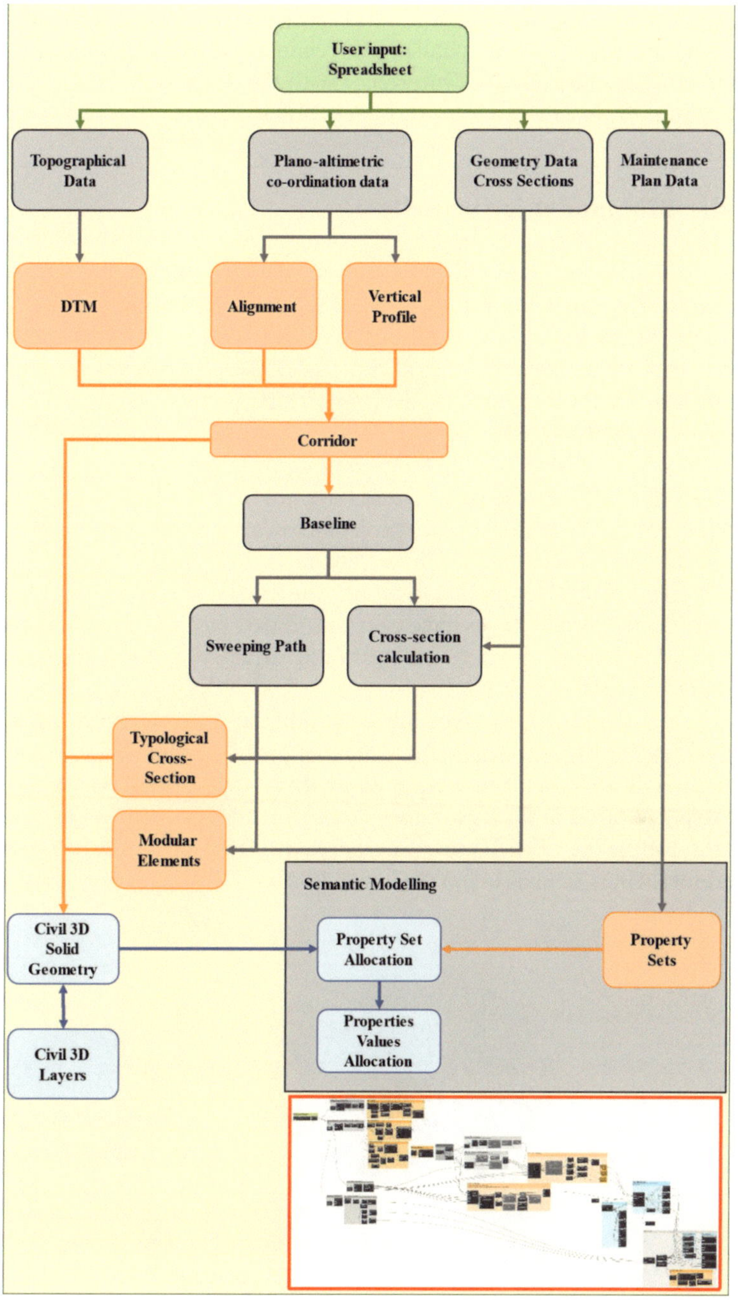

Figure 4.
Algorithm for automatic modelling of a road infrastructure.

extracts 'solid' objects to which it can associate 'property sets,' that is lists of properties that can take different formats (textual, numerical, formulas, images, etc.) and that constitute an enrichment to the semantic content of the digital information model.

Although Dynamo's default node library is well-stocked with functions and instructions, in some steps, such as creating property sets from scratch and assigning attributes to geometries, it was necessary to refer to libraries published by experienced users and the creation of custom nodes. This is possible using programming codes, such as Python and C#. The main challenge is related to the complexity of the Civil 3D API (Application Programming Interface) with the difficulty of having to navigate through hundreds of possible functions, instructions, constructors and methods.

5. Hydraulic-BIM modelling workflow

First, the design of hydraulic infrastructure in the BIM environment is evaluated for the authoring software Infraworks; it integrates BIM and GIS methodology, generating a workflow for moving data from one system to another without issues. This software tool allows for achieving almost definitive preliminary-level projects. The resulting files are interoperable with several software tools produced by the same manufacturer without data loss. A software tool that integrates GIS data from the manufacturer Esri with BIM data from the Autodesk manufacturer has a high potential. The knowledge of the natural and built environments of the context where to carry out a civil infrastructure project allows for timely and more informed decision-making, greater stakeholder involvement and faster approval processes. The second step is sizing the pipeline network; the third is verifying and controlling the pressure network. Both phases have been simplified by an innovative procedure within the I-BIM practice applied to the hydraulic infrastructures of the network under pressure; in particular, it has been combined with the software that allows the data population of the pipelines, a specific hydraulic calculation algorithm for network planning, verification and management. This tool procedure has been implemented because the Civil 3D software is not equipped with an integrated pressure network solver and requires other tools, usually available with additional payments. It was preferred to combine the Civil 3D software only with free tools, widely known in the sector; therefore, Civil 3D was integrated with EPANET and EpaCAD. **Figure 5** shows

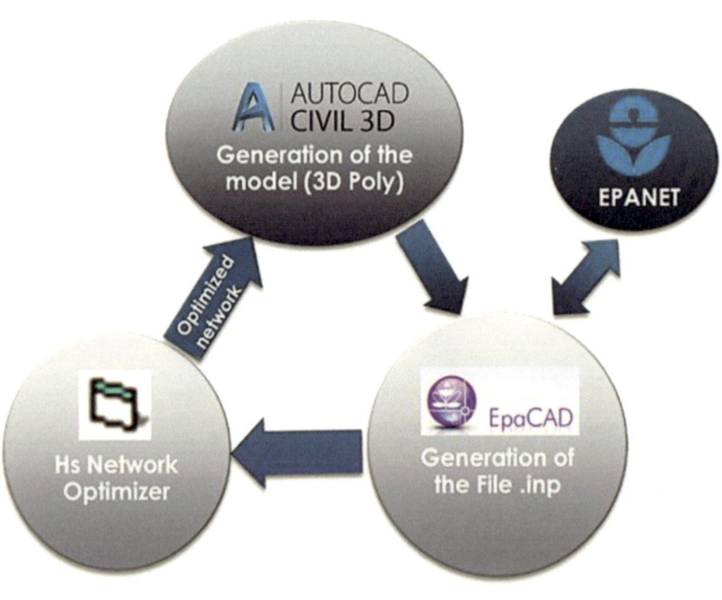

Figure 5.
Algorithm for automatic modelling of a pipeline network.

an innovative procedure for pressure water distribution networks. AutoCAD Civil 3D allows managing the horizontal-vertical and infrastructural information of an urban area. It allows automatically generating the 3D surfaces to represent the elevation profile of linear infrastructures. More specifically, using the 3DPOLY command, a closed polyline representing the water network (in this specific case study, two loops) can be drawn. In the 3D model, the network nodes are characterised by heights, defined by choosing the base surface of the vector cartography as the zero level. The preliminary network can be transformed using open-source software, such as EpaCAD (2009), to generate files with the .inp extension, readable through EPANET software. The .inp file can be edited by using HS network optimiser models. The objective function is minimised, accounting for costs and installation while respecting hydraulic and network constraints. The software tool is based on the Harmony search optimisation algorithm and interacts with EPANET to calculate pressures and flows in the water network. Finally, through AutoCAD Civil 3D, the final project of the water network is obtained by defining first the horizontal and then the vertical layout.

6. Conclusions

This research focuses on how design development and modelling of linear infrastructures can be carried out in a BIM environment. In particular, starting from the state of the art on BIM, a generally valid and proven workflow for the realisation of models with advanced software was described. The possibility of automating some if not all of the operations required for the realisation of a road BIM is a strong innovative element. The workflow described at the outset, while drastically reducing processing times compared to more traditional methods, is still time-consuming. The algorithm proposed by VIASTRATA reduces processing time to a few seconds, instead shifting the effort and commitment of professionals to the proper compilation of a specially designed spreadsheet. In fact, the main result of this work is precisely that of having found a methodology capable of drastically reducing modelling time in relation to a well-established framework for processing infrastructure BIM models. This opens up the possibility of expanding this approach to all transportation-related works and beyond. The algorithm was developed specifically for road infrastructure, but the principle is applicable to all linear infrastructure (pipelines, sewers, oil pipelines, railways, airport runways, etc.). In fact, with a few appropriate modifications, the level of generalisation of the algorithm would be such that it could be applied to all fields of linear infrastructure.

Research in the area of digitisation of infrastructure projects is in full swing. The need to adopt methodological approaches based on advanced digital technologies, such as BIM, is also supported by policy decisions, resulting from several years of experimentation and data collection regarding their effectiveness in technical-economic terms.

The potential of these tools is such that they can transform the way infrastructure projects are conceived and implemented. Future research directions based on the evolution of the algorithm presented in this chapter could involve integrating it with artificial intelligence (AI) systems for the evolution of generative modelling processes. In fact, the algorithm produces the model of a road infrastructure based on a rigid pattern of input data, while with AI it could autonomously learn how to respond to a wide variety of design needs.

Therefore, it is of paramount importance to continue exploring the possibilities offered by BIM and to increase its performance. Nevertheless, the algorithm developed in VPL described in this paper represents a possible application of a very important tool that has all the potential to expand the frontiers of BIM as a scientific methodology and as a technical-economic tool for the recovery of the construction and linear infrastructure sector.

Conflict of interest

The authors declare no conflict of interest.

Author details

Salvatore Antonio Biancardo[1,2*], Mattia Intignano[1,2], Francesco De Paola[1,2] and Gianluca Dell'Acqua[1,2]

1 Department of Civil, Construction and Environmental Engineering, University of Napoli Federico II, Napoli, Italy

2 VIASTRATA Spin-Off, University of Napoli Federico II, Napoli, Italy

*Address all correspondence to: salvatoreantonio.biancardo@unina.it

References

[1] Borrmann A, Konig M, Kock C, Beetz J. Building Information Modeling, Technology Foundations and Industry Practice. Cham: Springer; 2018. p. 584. DOI: 10.1007/978-3-319-92862-3

[2] Fröhlich C, Mettenleite M. Terrestrial laser scanning—New perspectives in 3D surveying. The International Archives of the Photogrammetry, Remote Sensing and Spatial Information Sciences. 2024;**36**:W2

[3] Stal C, Verbeurgt J, De Sloover L, De Wulf A. Assessment of handheld mobile terrestrial laser scanning for estimating tree parameters. Journal of Forestry Research. 2021;**32**:1503-1513. DOI: 10.1007/s11676-020-01214-7

[4] Achille C, Adami A, Chiarini S, Cremonesi S, Francesco F, Fregonese L, et al. UAV-based photogrammetry and integrated technologies for architectural applications—methodological strategies for the after-quake survey of vertical structures in Mantua (Italy). Sensors. 2015;**2015**(15):15520-15539

[5] Banfi F, Brumana R, Stanga C. Extended reality and informative models for the architectural heritage: From scan-to-BIM process to virtual augmented reality. Virtual Archaeology Review. 2019;**10**:14-30. DOI: 10.4995/var.2019.11923

[6] Brumana R, Tucci G, Lerma Garcia JL. Special section preface: Informative models and systems for virtual museums. Virtual Archaeology Review. 2019;**10**:ii-iii. DOI: 10.4995/var.2019.12357

[7] Baik A. From point cloud to Jeddah heritage BIM Nasit historical house – Case study. Digital Applications in Archaeology and Cultural Heritage. 2017;**4**:1-18. DOI: 10.1016/j.daach.2017.02.001

[8] Intignano M, Biancardo SA, Oreto C, Viscione N, Veropalumbo R, Russo F, et al. Stone paved road digital reproduction: A workflow. In: TRANSBALTICA XII: Transportation Science and Technology, Lecture Notes in Intelligent Transportation and Infrastructure. Cham: Springer; 2022. pp. 131-139. DOI: 10.1007/978-3-030-94774-3_13

[9] Bilov V, Goi V, Mamonov K, Tregub O, Levchenko O. Advantages of building information modeling (bim) during the operational life. Amazonia Investiga. 2023;**12**(68):346-363

[10] Moon HS, Kim CY, Cho GH, Moon JS, Ju KB. Interoperability verification using BIM case models of road project. Journal of KIBIM. 2015;**5**(1):44-53

[11] Biancardo SA, Gesualdi M, Savastano D, Intignano M, Henke I, Pagliara F. An innovative framework for integrating cost-benefit analysis (CBA) within building information modeling (BIM). Socio-Economic Planning Sciences. 2023;**85**:101495. DOI: 10.1016/j.seps.2022.101495

[12] Biancardo SA, Intignano M, Viscione N, Guerra De Oliveira S, Tibaut A. Procedural modeling-based BIM approach for railway design. Journal of Advanced Transportation. 2021;**2021**:1-17

[13] Lee SS, Kim KT, Tanoli WA, Seo JW. Flexible 3D model partitioning system for nD-based BIM implementation of alignment-based civil infrastructure.

Journal of Management in Engineering. 2020;**36**(1):04019037

[14] Chong H, Lopez R, Wang J, Wang X, Zhao Z. Comparative analysis on the adoption and use of BIM in road infrastructure projects. Journal of Management in Engineering. 2016;**32**:1-13

[15] Sankaran B, O'Brien W, Goodrum P, Khwaja N, Leite F, Johnson J. Civil integrated management for highway infrastructure: Case studies and lessons learned. Transportation Research Record: Journal of the Transportation Research Board. 2016;**2573**:10-17

[16] Abdelwahab HT. Intelligent design (4D, 5D and beyond) for road design and construction projects: Two case studies. International Road Federation (IRF) Examiner. 2018;**12**:21-25

[17] Abbondati F, Biancardo SA, Sicignano G, Guerra de Olivera S, Tibaut A, Dell'Acqua G. BIM parametric modeling of a railway underpass. Ingegneria Ferroviaria. 2020;**6**:443-459

[18] Salzano A, Intignano M, Mottola C, Biancardo SA, Nicolella M, Dell'Acqua G. Systematic literature review of open infrastructure BIM. Buildings. 2023;**13**(7):1593. DOI: 10.3390/buildings13071593

[19] Biancardo SA, Intignano M, Abbondati F, Abramović B, Dell'Acqua G. Horizontal building information modeling: The Croatian railway Gradec-Sveti Ivan Žabno case study. Ingegneria Ferroviaria. 2021;**76**(12):979-994, Cited 3 times

[20] Biancardo SA, Intignano M, Veropalumbo R, Martinelli R, Calvanese V, Autelitano F, et al. BIM approach for stone pavements in archaeological sites: The case study of Vicolo Dei Balconi of Pompeii. Transportation Research Interdisciplinary Perspectives. 2023;**17**:100755. DOI: 10.1016/j.trip.2023.100755

[21] Biancardo SA, Viscione N, Cerbone A, Dessì E Jr. BIM-based design for road infrastructure: A critical focus on modeling guardrails and retaining walls. Infrastructures. 2020;**5**:59

[22] Tschickardt T, Krause D. BIM in highway construction using the example project availability model A 10/A 24. Bautechnik. 2019;**96**(3):259-268

[23] Bosurgi G, Celauro C, Pellegrino O, Rustica N, Sollazzo G. The BIM (building information modeling)-based approach for road pavement maintenance. Lecture Notes in Civil Engineering. 2020;**48**:480-490

[24] Jing W, Hao G, Li C, Wei W, Cheng J. BIM application approach on highway maintenance and management. In: Proceedings of the 19th COTA International Conference of Transportation Professionals; 6-8 July 2019; Nanjing. China. Reston, VA: ASCE; 2016. pp. 776-786

[25] Biancardo SA, Intignano M, Pires DM, Abbondati F, Dell'Acqua G. Heritage BIM approach for roman pavements. European Transport - Trasporti Europei. 2023;**91**:8. DOI: 10.48295/ET.2023.91.8

[26] Patel K, Ruparathna R. Life cycle sustainability assessment of road infrastructure: A building information modeling-(BIM) based approach. International Journal of Construction Management. 2023;**23**(11):1837-1846. DOI: 10.1080/15623599.2021.2017113

[27] Gan VJL, Liu T, Li K. Integrated BIM and VR for interactive aerodynamic design and wind comfort analysis of modular buildings. Buildings. 2022;**12**:333. DOI: 10.3390/buildings12030333

[28] Deng Y, Cheng JCP, Anumba C. A framework for 3D traffic noise mapping using data from BIM and GIS integration. Structure and Infrastructure Engineering. 2016;**12**(10):1267-1280. DOI: 10.1080/15732479.2015.1110603

[29] Wang J, Hou L, Wu P. BIM-supported tunnel light environment evaluation: A case study on Shanghai Chenxiang road tunnel project. In: Proceedings of the 16th International Conference on Computing in Civil and Building Engineering (ICCCBE 2016). Osaka, JPN: Osaka University; 2016. pp. 1023-1030. Available from: http://www.see.eng.osaka-u.ac.jp/seeit/icccbe2016/pages/conference5.html

[30] Cantisani G, Correa Panesso JD, Del Serrone G, Di Mascio P, Gentile G, Loprencipe G, et al. Re-design of a road node with 7D BIM: Geometrical, environmental and microsimulation approaches to implement a benefit-cost analysis between alternatives. Automation in Construction. 2022;**135**:104133. DOI: 10.1016/j.autcon.2022.104133

[31] Shadram F, Johansson TD, Lu W, Schade J, Olofsson T. An integrated BIM-based framework for minimizing embodied energy during building design. Energy and Buildings. 2016;**128**:592-604. DOI: 10.1016/j.enbuild.2016.07.007

[32] Tang F, Ma T, Zhang J, Guan Y, Chen L. Integrating three-dimensional road design and pavement structure analysis based on BIM. Automation in Construction. 2020;**113**:103152. DOI: 10.1016/j.autcon.2020.103152

[33] Collao J, Lozano-Galant F, Lozano-Galant J. A., Turmo, J. BIM visual programming tools applications in infrastructure projects: A state-of-the-art review. Applied Sciences. 2021;**11**(18):8343

[34] VIASTRATA spin-off. Available from: https://www. viastrata.unina.it/

Chapter 5

3D Printing in Highway Construction, Opportunities and Challenges

Mohammadsoroush Tafazzoli, Fatemeh Naeijian and Syeda Farwa Narjis Naqvi

Abstract

This chapter explores the transformative impact of 3D printing technology within highway construction, offering a detailed examination of both its opportunities and challenges. It outlines the advantages of adopting 3D printing particularly to construct highway bridges, including enhanced safety, boosted productivity, reduced labor demand, fostering of innovative design, efficient use of materials, savings on transportation, fuel conservation, accelerated construction speed, and waste reduction, thus revolutionizing the future of infrastructure development. The chapter argues for more research into material procurement, structural concerns, and the environmental benefits of 3D-printed highway bridges. By overcoming these obstacles, 3D printing could lead to more sustainable, efficient, and innovative ramps and bridges, marking a significant shift towards high-speed construction and improved constructability, even under extreme conditions. The conclusion underscores the critical role of collaboration among stakeholders to navigate the challenges and realize the potential advantages of 3D printing in transforming road construction for future generations.

Keywords: 3D-printing, bridge construction, manufacturing, automation, smart roads, scalability, digital fabrication

1. Introduction

1.1 Definition

3D printing is a process where materials are systematically deposited layer by layer through a print head, nozzle, or other advanced technologies to construct physical structures directly from 3D models [1]. This method has brought significant advancements to industries including manufacturing, medical, and food, and has recently started to impact the construction sector [2]. Construction 3D printing is an approach that allows for the fabrication of different types of concrete structures in a manner that is both efficient and sustainable [3]. This technology offers a level of customization and complexity in design that was not possible with traditional methods.

1.2 How 3D-printed structures are built

The process of building 3D-printed structures begins with the creation of a specialized materials mixture, engineered to meet the unique structure and architectural demands. Essential qualities of this mixture include (1) ease of extrusion through the printing nozzle without clogging, (2) ability to flow smoothly and form even layers, and (3) high capacity to maintain shape and support the weight of subsequent layers without collapsing. Using thinner layers improves surface quality, though it requires more time [4].

The foundation of most 3D printing mortars is a blend of cement-based binders for strength, superplasticizers to improve flow, polypropylene fibers to prevent cracks and enhance mechanical strength, sand as a basic aggregate, and water, which is crucial for the hydration of cement. These elements are mixed into a consistent paste, which is then layered incrementally to construct the structure. Aware of the environmental concerns associated with cement production, notably its significant CO_2 emissions [5], there is an increasing effort to incorporate supplementary cementitious materials (SCMs) such as silica fume, limestone filler, metakaolin, fly ash, blast-furnace slag, nano-silica, and sugarcane bagasse ash into 3D printing mixtures [6].

The construction of 3D-printed bridges could be executed by a movable printer that travels along the designated path, extruding the prepared mixture through a nozzle to deposit material in successive layers. This method allows for the fabrication of complex designs and the integration of features such as drainage systems or sensors with unparalleled precision and efficiency. Controlled digitally, this process enables real-time adjustments and optimizations to cater to the specific requirements and design of the bridge [7], showcasing how 3D printing technology tailors advanced solutions to the evolving needs of bridge construction. **Figure 1** shows the major steps in building 3D-printed pavements.

1.3 Potentials

The adoption of 3D printing in roadway construction presents a myriad of opportunities and challenges that could significantly impact efficiency, sustainability, and innovation in infrastructure development. 3D printing offers significant advantages in the construction industry, including improved efficiency in terms of cost and environmental impact. It also enables the widespread customization of designs [8]. **Figure 2**

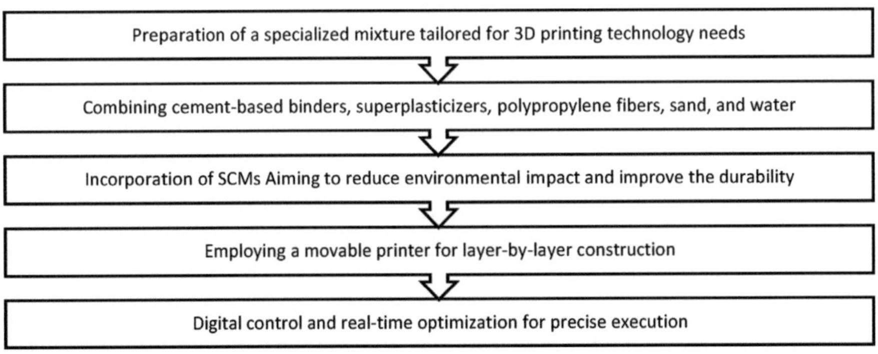

Figure 1.
Major steps in building 3D-printed pavements.

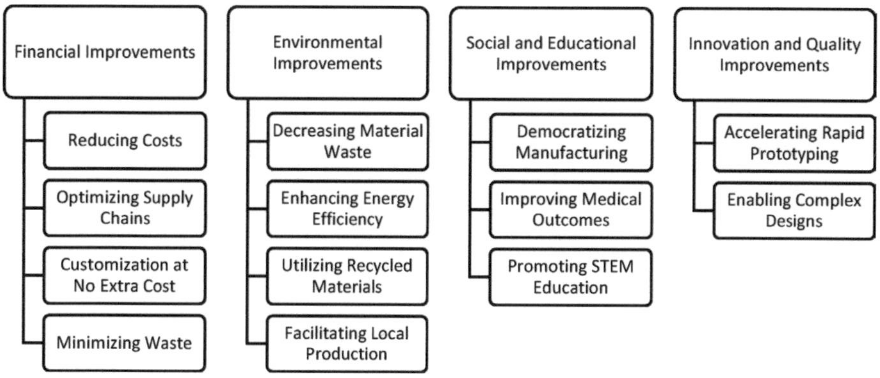

Figure 2.
The potential of adopting 3D-printing in multiple categories.

shows these potentials in multiple categories. This potential allows for the tailoring of esthetics, functionality, and structural integrity to meet specific project needs. This chapter aims to explore the emerging role of 3D printing in highway bridge construction, delving into its potential benefits and the hurdles that must be overcome to realize its full potential.

1.4 The demand for adopting new technologies in construction

The construction sector, particularly in the realm of road construction, stands at a pivotal juncture where embracing innovation could significantly alter its landscape for the better. Historically, the construction industry has a reputation

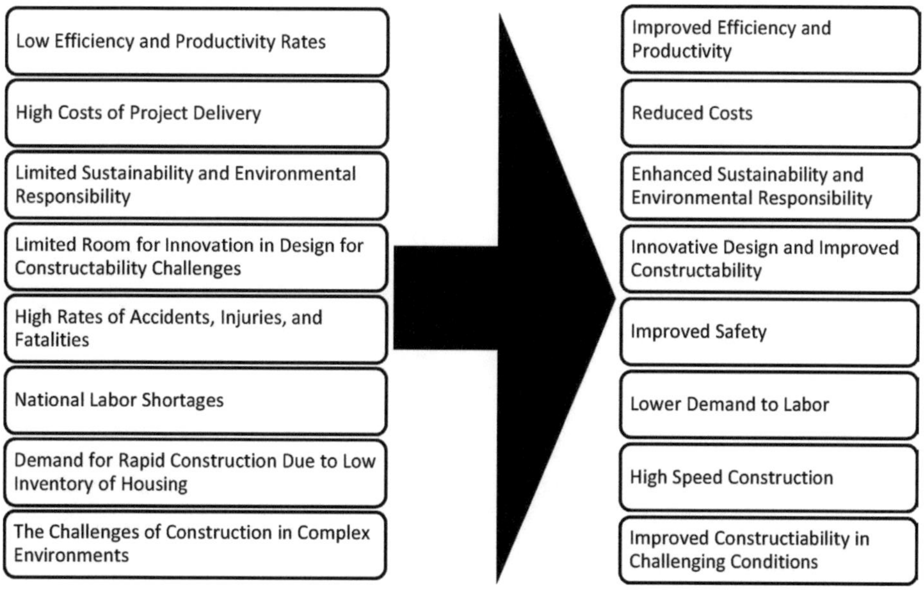

Figure 3.
Potential outcomes of adapting new technologies in construction.

for inefficiencies [9], notably due to the lack of standardization [10], leading to substantial losses in the workflow. This inefficiency is highlighted by the sector's meager average global growth in labor productivity, which has lingered at around 1% per annum over the past two decades, starkly contrasted by the 3.6% growth seen in manufacturing [1]. The root of this disparity lies not only in the sector's reluctance to adopt new technologies and modern management but also in its minimal investment in research and development, especially when compared to the manufacturing industry [11]. **Figure 3** illustrates why the adoption of new technologies seems to be a must in upcoming years.

2. Background

Beyond constructing entire buildings, 3D Concrete Printing (3DCP) has seen successful applications across various sectors within the construction industry. It has been utilized to fabricate pedestrian bridges (see **Figure 4**), perform in situ repair works in locations that are challenging and hazardous for human access, and rapidly erect disaster relief shelters and military bunkers [12]. Additionally, 3D printing is employed in the creation of metal frames, both structural and non-structural elements with intricate geometries, and the production of molds for load-bearing components [12]. It also plays a crucial role in the reproduction of components for historical buildings, showcasing its versatility and potential to transform traditional construction methodologies.

3D-printed bridges have moved from being merely a concept to actual implementation in several pioneering projects around the world, marking a significant step towards innovation in civil engineering and infrastructure development. These projects indicate the practical applications and potential benefits of using 3D printing technology in bridge construction. In 2018, the The Netherlands opened the first-ever bridge made of 3D-printed concrete specifically for cyclists, showcasing the potential of 3D printing technology in infrastructure development. This project paved the way for considering similar technology for road and ramp construction.

Figure 4.
3D-printed pedestrian bridge in Amsterdam, Netherlands, built in 2021.

3. Current applications of 3D printing in roadway construction

3.1 Automated road repair with 3D printing

Researchers are exploring the use of drones equipped with 3D printers for automated road repairs. These drones could potentially scan road surfaces continuously, identifying and filling small cracks early to prevent them from worsening. This approach aims to make road maintenance more affordable and efficient. A team at University College London (UCL) has developed the inaugural 3D printer of its kind. Simultaneously, their counterparts at Leeds University have been working on a pioneering drone designed to transport this printer to sites requiring repairs. This is shown in **Figure 5**.

3.2 Asphalt 3D printer for safer roads

Researchers at Monash University and Chang'an University have created an innovative 3D printer for asphalt that can autonomously identify and repair cracks in pavements and roads. This device incorporates a camera and a specialized FDM 3D printer, which is enhanced with a unique image-processing algorithm, allowing it to independently fill in cracks and enhance the safety of roads. **Figure 6** shows this device [14].

3.3 Development of asphalt 3D printers

Numerous efforts have been made to create asphalt 3D printers over recent years, yet these innovations have not reached a level of efficiency that justifies their practical use on large scales. In 2015, Advanced Paving Technologies initiated a Kickstarter campaign to support the development of its innovative Asphalt Paving Machine, integrating Lidar 3D scanning technology [15]. This technology was designed to identify imminent road fractures for repair in a manner that is quicker, more cost-effective, and cleaner than traditional manual methods.

Following this, the University of Leeds embarked on a project to engineer tarmac 3D printing robots. These robots leverage data captured by Unmanned Ground Vehicles (UGVs) to locate and mend potholes. There's potential for these devices to be adapted for Unmanned Aerial Vehicles (UAVs) in the future, enabling the repair

Figure 5.
Leeds University's repair drone [13].

Figure 6.
Automated 3D printing for pavement repair in concrete and asphalt.

of difficult-to-reach areas. Likewise, the University College London (UCL) team has introduced a specialized asphalt 3D printer tailored for pothole repairs. This mobile device allows for the precise management of 3D-printed asphalt parameters, enhancing the efficiency of the extrusion process.

4. Opportunities of adopting 3D printing in bridge construction

The building sector is poised for a transformative change due to the emergence of 3D printing technology. This shift is expected to revolutionize standard procedures by enhancing efficiency in terms of time, costs, and materials. The subsequent sections will examine the advantages of incorporating 3D printing in construction. Additionally, this discussion will cover how 3D printing might influence the future of constructing highways.

4.1 Improved safety

The construction industry has long been recognized for its elevated rates of fatalities, injuries, and illnesses, marking it as one of the more hazardous fields compared to other sectors [16]. This high-risk environment calls for innovative solutions to enhance safety and efficiency on construction sites [17]. The integration of 3D printing technology, especially when combined with Building Information Modeling (BIM), presents a promising avenue to address these safety concerns while also streamlining construction processes.

3D printing in construction offers the potential to significantly reduce the time workers spend on-site [18], thereby minimizing their exposure to dangerous conditions. By prefabricating parts and components off-site in controlled environments, the technology can lower the risk of accidents associated with traditional construction methods. This approach not only improves worker safety but also boosts the overall efficiency of construction projects [18].

The capability of 3D printing to work under harsh conditions also opens up new possibilities for construction in environments that are inherently dangerous or difficult to access [12]. For example, in areas prone to extreme weather, 3D printers can operate in enclosed spaces, shielded from the elements, to produce components that are later assembled on-site quickly and safely. Similarly, for projects in remote or hazardous locations, such as offshore platforms or disaster zones, 3D printing can enable the construction of structures with minimal human intervention, significantly reducing risk. **Figure 7** shows how adopting 3D printing can contribute to construction safety.

4.2 Improved productivity

The construction industry, despite being a significant economic contributor, has long been marked by low productivity, characterized by minimal technological advances and limited automation [10]. 3D printing, with its capacity for automation, presents a substantial opportunity to enhance efficiency across the construction process. Unlike traditional methods, 3D printing can streamline the construction workflow, significantly reducing losses that typically occur between processes and sub-processes. This improvement is largely due to the technology's reliance on digital models, which facilitate a more standardized approach to construction, minimizing the discrepancies and inefficiencies that plague conventional methods.

Labor productivity in construction has languished at a global average growth rate of only 1% per year over the past two decades [19]. By contrast, the manufacturing industry, which has embraced automation and technological innovation more readily, has seen a 3.6% growth in the same period [19]. 3D printing can automate the production of bridge components, reducing the reliance on manual labor and mitigating the impact of human error [20].

4.3 Reduced demand for labor

The construction industry is currently facing a national shortage of laborers [21], a challenge that is exacerbating the pressures of meeting growing infrastructural demands. 3D printing technology emerges as a transformative solution to this issue, promising not only to reduce the sector's dependence on skilled labor but also to refine the overall construction process. By automating significant portions of construction, 3D printing technology can mitigate the labor shortage problem by requiring fewer workers on-site. This reduction in manual labor is not merely a response to workforce shortages; it also contributes to enhancing the safety of construction sites as was discussed above.

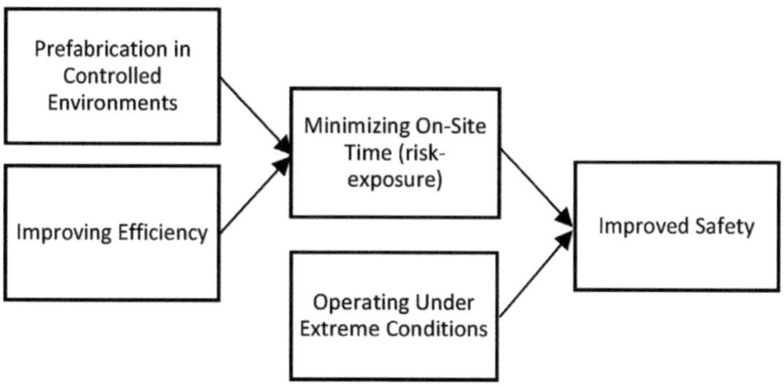

Figure 7.
How 3D printing contributes to construction safety.

4.4 Efficient material use

3D printing in construction facilitates the use of a wide variety of materials, including those that traditional construction methods might not easily accommodate. Notably, this includes recycled plastics and other sustainable alternatives [3], thereby extending the lifecycle of materials that would otherwise contribute to landfill waste. Using these materials in the construction of highway bridges not only lessens the environmental effects of these endeavors but also supports a circular economy. This approach encourages the reuse and recycling of materials, diminishing the need for new resources and lessening the carbon emissions linked to their extraction and processing.

Moreover, the repair and maintenance of bridges built using 3D printing can be more efficient and less resource-intensive. The digital models used in 3D printing allow for precise replication or modification of sections that require repair, facilitating quicker interventions with minimal waste [22]. This can ensure the longevity and durability of infrastructure and also support sustainable management practices throughout the lifecycle of the bridge.

4.5 Transportation savings

Adopting 3D printing in construction offers a compelling avenue for reducing the extensive demand for transportation associated with traditional construction methods. This innovative approach diminishes the need for transporting a vast array of materials [20]. Traditional construction often requires the delivery of diverse materials from various locations, contributing to a high volume of transportation activities. In contrast, 3D printing streamlines this process by utilizing a single or a limited number of materials, thus substantially lowering transportation needs.

The reliance on heavy equipment such as pumps, cranes, compressors, generators and lifts, and is drastically reduced with 3D printing techniques [20]. These machines typically necessitate frequent onsite and offsite movement, including transportation to the construction site at the beginning of a project and removal upon its completion, as well as possible transfers between different project sites. The minimized use of such equipment in 3D printing lessens transportation demand and contributes to reducing the carbon footprint associated with these movements.

Another ripple effect of reduced demand for transportation is decreased labor and staff requirements. Traditional construction methods often involve large teams of workers who commute to and from the construction site to operate the equipment, leading to numerous work-related trips. The automation and efficiency of 3D printing mean fewer personnel are needed on-site, thereby reducing the frequency of these commutes and any associated onsite transportation of materials and equipment by the workforce. Certain 3D printing techniques, such as D-Shape, advocate for the use of local materials in the construction process [23]. This approach can lead to considerable savings in transportation energy, as it eliminates the need to source materials from distant locations. By utilizing locally available resources, the overall energy consumption and emissions associated with transporting construction materials over long distances could be markedly reduced.

4.6 Fuel savings

Adopting 3D printing in construction minimizes the need for traditional construction equipment and labor and significantly shortens construction times, leading to a substantial reduction in the energy required for manufacturing [24]. Conventionally, a wide array of equipment and power tools are employed in construction, many of which rely on Energy sources derived from fossil fuels, including electric power, natural gas, and diesel fuel. The shift towards 3D printing and the use of robotic systems, which primarily run on electricity, presents an opportunity to utilize local renewable energy sources, further reducing the carbon footprint of construction projects.

4.7 Reduced downtime and delays

The highly mechanized and automated nature of 3D printing in construction means less reliance on human labor. This not only streamlines the construction process but also mitigates the risk of delays associated with labor strikes, economic downturns, and adverse weather conditions [25]. Additionally, the precision of 3D printing minimizes the likelihood of rework or damage, both of which are common sources of additional energy expenditure in traditional construction methods. The adoption of construction 3D printing could lead to a 63% reduction in costs associated with formwork [2]. Similarly, Malaeb et al. [26] estimate that utilizing 3D printing in construction could decrease the total cost of construction by 35–60% by eliminating the need for formwork.

4.8 Decrease in pollutants and disturbances

The acceleration of construction processes through 3D printing is achieved by its ability to fabricate components directly from digital designs, eliminating many of the intermediate steps required in conventional construction. This direct fabrication process enables continuous construction activity without the need for setting, curing, and other time-consuming phases associated with traditional materials like concrete. Additionally, 3D printing enables the creation of intricate designs quickly and accurately, tasks that would typically be difficult and time-consuming with traditional manual techniques. The rapid pace of construction made possible by 3D printing also plays a crucial role in enhancing sustainability. Fast-track construction reduces the overall time machinery and equipment are in use on-site, leading to lower emissions and less environmental disturbance. Additionally, the expedited construction process decreases the duration of potential disruptions to natural habitats and local communities, further aligning projects with sustainability objectives [3]. Therefore, the accelerated construction pace achieved through 3D printing significantly mitigates common construction-related nuisances such as smoke, dust, and excessive noise. This reduction in pollutants and disturbances greatly enhances the health and comfort of neighboring residents, providing a cleaner and quieter environment during the construction phase. By minimizing these typical issues, 3D-printed construction offers a more neighbor-friendly approach to building projects.

4.9 Quick repairs

The rapid construction capabilities of 3D printing have profound implications for repair and maintenance operations. Concrete road surfaces often crack due to drying shrinkage, and to control where these cracks occur, cracks are deliberately induced at specific locations. However, these predetermined joints become focal points for vehicle impact loads, leading to significant palm-sized spall damage. Traditional repair methods using cast-in-place concrete not only prolong road closures during the curing period but also cause substantial indirect losses due to traffic disruption [27].

In a study by Yeon et al. [27] an innovative repair technique using a pre-prepared 3D concrete patch was introduced, which significantly reduces the road closure time to just 2 hours. According to the US Department of Transportation, this new method could decrease indirect repair costs from $140,000 to $1700. The study confirms that the patch, which can handle up to 91% of the shear load strength of traditional methods, may also offer superior structural stability due to the friction between the patch and the existing surface [27].

The ability to quickly produce and replace damaged sections of infrastructure can lead to more efficient and less disruptive maintenance activities, ensuring that highways remain in optimal condition with minimal impact on traffic flow. This aspect of high-speed construction enhances the overall sustainability and reliability of infrastructure, demonstrating the far-reaching benefits of integrating 3D printing into highway construction projects.

4.10 Constructability in extreme conditions

Where extreme conditions prevail, such as remote locations, harsh climates, or disaster-stricken areas, traditional construction methods grapple with heightened logistical complexities, increased costs, and heightened risks to workforce safety. These challenges are amplified for bridge construction projects with labor-intensive requirements due to limited resources and a tendency to integrate rather than innovate technological solutions.

3D printing technology offers a unique proposition to address these challenges by facilitating the construction of structures directly on-site with minimal labor requirements. This is particularly advantageous in extreme environments where the risks associated with traditional construction methods are exacerbated. The technology's capacity for customization and its ability to utilize a variety of materials, including those locally sourced or specifically engineered for resilience, allows for the construction of structures optimized for the unique demands of extreme conditions [8].

4.11 Waste reduction

The adoption of 3D printing in bridge construction offers a significant opportunity for waste reduction, aligning with the global trends towards digitization, the adoption of new materials, and energy efficiency. By utilizing digital models, 3D printing enables the precise control of material usage, thus minimizing waste generated during the construction process. This precision ensures that materials are used efficiently, contributing to the construction of structures that are not only designed to meet specific environmental challenges but are also energy-efficient [23]. This approach not only addresses immediate constructability concerns but also aligns with broader sustainability and efficiency objectives, making a valuable contribution to reducing the environmental footprint of construction projects.

However, the sustainability of 3D printing in construction also depends on the management of waste generated at the end of a structure's lifecycle, particularly the debris resulting from the demolition of 3D printed segments. There is a need for further research to explore how this debris can be safely disposed of and to assess any environmental concerns associated with its disposal. Investigating the recyclability of 3D printed materials and the potential for repurposing or reusing demolition waste could provide insights into full lifecycle sustainability. Addressing these questions is crucial for ensuring that 3D printing technology not only reduces waste during construction but also promotes environmental stewardship throughout the entire lifespan of constructed assets.

4.12 Replacing steel needed for reinforcement with low-cost alternatives

The use of 3D printing in constructing bridges offers a promising option for substituting traditional steel reinforcement with more cost-effective materials. This is achieved through the Extrusion Deposition (ED) method of 3D printing, which employs polymers like acrylonitrile butadiene styrene (ABS), polycarbonate (PC), polylactide (PLA), polyamide (PA), and their blends [28]. These materials offer the benefits of low melting points and low distortion, alongside anisotropic mechanical properties. However, their inherent lack of mechanical strength has been a limiting factor, a challenge that recent developments in polymer matrix composites aim to overcome. By reinforcing these polymers with particles, fibers, and nanomaterials, it is now possible to enhance their strength, making them suitable alternatives to steel reinforcement in bridge construction [29].

This holds particular significance considering that around one-fourth of all steel manufactured is utilized in construction, with the average estimated usage of this steel falling below 50% of its available capacity [30]. The high cost of steel has been a significant factor in the overall expenses associated with bridge construction. Replacing steel with these innovative, lower-cost polymer composites could lead to substantial cost reductions in construction projects. This not only has the potential to make bridge construction more economically feasible but also opens up opportunities for broader application and innovation in the field. Ongoing research is crucial to further improving the tensile strengths of 3D printed bridges, ensuring they can meet or exceed the performance standards set by traditional materials.

4.13 Integration with Building Information Modeling

The integration of Building Information Modeling (BIM) with 3D printing presents a revolutionary approach to construction, particularly in the creation of highly optimized construction components. Building Information Modeling (BIM) provides a digital model that encapsulates both the physical and functional aspects of a facility, creating a detailed information resource for the building process. When integrated with 3D printing technology, this combination enhances the efficiency, accuracy, and smoothness of the construction workflow, from initial design to final implementation [31].

When the optimized BIM designs are fed into a 3D printer, the technology allows for the direct fabrication of components as envisioned in the BIM models. 3D printing can accurately replicate the complex shapes and intricate details designed in BIM, with a level of precision that traditional construction methods struggle to achieve. This capability is particularly beneficial for custom or unique elements of a project, where 3D printing can produce components that perfectly match the digital design with minimal waste.

The combination of BIM and 3D printing supports the pursuit of sustainable construction practices. By precisely calculating the amount of material needed and optimizing the design for material efficiency, the integrated approach significantly reduces waste. Additionally, the ability to print components on-demand and on-site can lower transportation costs and emissions associated with moving materials from off-site locations [18].

5. Challenges facing 3D printing in highway construction

Although 3D printing offers significant benefits, its implementation in the construction of highways and bridges faces several challenges. The industry's hesitance to embrace new technologies is exacerbated by issues such as a shortage of workers trained to operate sophisticated equipment, limited collaboration with researchers, and challenges in synchronizing commercial objectives with scientific research. Furthermore, Small and Medium-sized Enterprises (SMEs), which make up a large part of the construction sector, typically focus on meeting labor-intensive needs rather than investing in technological innovation, opting instead to adopt external technological developments into their operations.

To overcome these challenges and harness the potential of 3D printing, the construction sector, particularly firms involved in bridge construction, must strategically position themselves on the side of innovation. This involves not only a shift in mindset but also a commitment to investing in R&D and fostering stronger collaborations with the research community. By embracing digital technologies, adopting new materials and processes, and focusing on energy efficiency, construction players can enhance their competitiveness and capitalize on the worldwide growth trend.

Identifying these challenges serves as a critical first step towards their resolution, paving the way for the more extensive utilization of 3D printing in construction. In the sections that follow, we will delve into some of these challenges, discussing their implications and exploring potential strategies for overcoming them, to fully harness the transformative power of 3D printing in the construction industry.

5.1 Structural concerns

The integration of 3D-printed elements into construction projects introduces a series of structural considerations that must be meticulously managed to ensure the reliability and safety of the resulting structures. Surface roughness, for instance, can significantly impact the ultimate tensile strength and rupture of 3D printed components, while dimensional and geometric inaccuracies can affect stress measurements derived from tensile tests [32]. Research into the structural performance of 3D printed metal elements for construction purposes has been notably scarce. Additionally, the integrity of the struts is crucial for maintaining the properties of micro strut blocks, with the microstructure and inclusions in the parent material potentially affecting all mechanical properties [31]. Although post-processing techniques such as chemical etching, electrochemical polishing, or heat treatment can enhance the quality of 3D printed micro-lattices, these methods invariably increase the overall cost [31]. Achieving optimal structural integrity in-built parts necessitates a deep understanding of several key process parameters, including scanning speed, layer thickness, local geometry, and the part temperature at the fusing point [31].

Adopting 3D printing in bridge construction can lead to uneven surfaces due to the inherent layering effect of the technology, which sometimes results in voids between layers [33]. This challenge is highlighted in existing literature, suggesting that the careful selection of layer thickness can mitigate the uneven surface effect. However, opting for thinner layers to improve surface smoothness increases the total printing time required to complete the structure [33]. To address this, tool path optimization is crucial, as identified by Jin et al. [32]. This process involves balancing the build time, the time gap between printing layers, and the desired surface finish as key variables [33].

5.2 Material procurement

The procurement of specific materials for 3D printing in construction presents unique challenges due to the complex nature of the materials required for the printing process. Developing materials that are compatible with 3D printing technologies demands a meticulous combination of various components to meet the structural and functional needs of construction applications. The advancement of 3D printing in the construction and building sector depends on improving materials to meet the unique needs of different applications. Enhancing materials by making them lightweight, thermally insulated, self-healing, and self-sensing adds complexity. However, this also substantially increases the utility and worth of the printed structures [33].

5.3 Technical limitations and challenges of reinforcing 3D-printed concrete

Technical limitations currently pose a significant hurdle. The size of 3D printers and the speed at which they can produce materials are constraints that impact the feasibility of using this technology for large-scale projects like highway construction. Advances in 3D printing technology are required to increase efficiency and scalability to levels that are competitive with traditional construction methods.

One of the significant challenges that hinder the adoption of 3D printing in bridge construction is the difficulty associated with reinforcing 3D printed concrete. Concrete, while strong under compression, has low tensile strength and ductility, making reinforcement essential for many construction applications, including bridges, to ensure their durability and resilience. Traditionally, this reinforcement has been achieved through the addition of steel bars or meshes within the concrete. However, the integration of such reinforcements is not straightforward in the context of 3D printing.

As demonstrated by Khoshnevis et al. [33] current methods for incorporating reinforcement into 3D-printed concrete, such as embedding reinforcements or post-tensioning reinforcement bars, require manual intervention. This manual insertion process complicates the 3D printing workflow, potentially diminishing the efficiency gains offered by the technology. Furthermore, designing structures with straight hollow voids to accommodate post-tensioning reinforcement bars could significantly restrict the architectural design freedom which is one of the key advantages of 3D printing.

An innovative solution, such as attaching a steel extrusion gun akin to a staple gun at the back of the nozzle, has been proposed to introduce tensile strength enhancements in the vertical direction [33]. Yet, this approach introduces its own set of challenges, particularly in controlling the force with which the steel staple penetrates the filament. Too much force could damage or destroy the fresh concrete, while insufficient force might result in inadequate penetration, thereby failing to provide the needed reinforcement [33].

Additionally, the incorporation of fibers into the concrete mix has been suggested as a means to improve ductility. However, the effectiveness of fibers and their compatibility with 3D printing processes in sufficiently reinforcing concrete for bridge construction needs further exploration [33].

5.4 Lack of manufacturing guidelines

One of the primary hurdles is the current lack of standardized manufacturing guidelines and practices, which results in a wide variability of outcomes from the same 3D CAD files across different additive manufacturing methods [4]. This inconsistency, advantageous for equipment makers by creating vendor dependency, significantly impedes the wider integration of additive manufacturing methods in the construction industry.

The absence of standardization spans several critical areas including materials, processes, calibration, and testing [34]. The construction industry's move towards 3D printing for building infrastructure such as concrete bridges, houses, and even high rises introduces a pressing need for uniform standards that can ensure consistency and reliability in the produced structures. Variability not only across builds but also between machines using the same process necessitates the development of new quality assurance measures [34]. These measures are essential to guarantee that the constructed parts meet the necessary requirements, a concern that becomes increasingly significant as we move towards structures subjected to the stresses and strains of everyday use, like bridges.

Unlike the traditional high-value sectors where 3D printing has been previously explored, bridge construction demands materials and construction techniques that can withstand heavy, continuous use and exposure to various environmental conditions. The challenges related to variability in material properties and geometry, inherent to the manufacturing technique, are compounded by the broader concerns of climate change, rapid urbanization, and resource scarcity [4]. These factors underscore the urgent need for more efficient, sustainable construction techniques [4].

Moreover, as the capabilities of additive manufacturing continue to evolve, many of the current technological limitations may be alleviated. However, this evolution brings its own set of challenges, including the need for ongoing adaptation of standards and quality assurance measures to keep pace with the advancements in manufacturing capabilities. The potential of 3D printing in revolutionizing bridge construction is immense, offering opportunities for increased efficiency, sustainability, and innovation. Yet, realizing this potential hinges on overcoming these significant challenges, requiring concerted efforts from industry stakeholders, regulatory bodies, and research institutions to establish a robust framework for the application of additive manufacturing in the construction sector.

5.5 Limited material technology

The limited technology of materials significantly hampers the adoption of 3D printing in bridge construction due to several key challenges that need to be addressed to meet the sector's unique demands. Bridge construction materials must possess specific properties such as high durability, resistance to a wide range of weather conditions, and the capacity to support heavy loads. Developing materials that not only meet these stringent criteria but are also compatible with 3D printing technology is an intricate process. The challenge lies in balancing the physical and mechanical properties required for bridge construction with the material characteristics needed for effective 3D printing, such as flowability, setting times, and adhesion between layers.

Ongoing research and development efforts aim to innovate and refine materials suitable for this purpose. However, the path to identifying and testing materials that can reliably fulfill the requirements of both 3D printing technology and bridge construction is complex. The process involves extensive experimentation to formulate materials that harden quickly enough to support successive layers without compromising the time needed to adjust or modify the print. Additionally, these materials must maintain their integrity and performance over time despite exposure to environmental stressors such as temperature fluctuations, moisture, and physical wear and tear.

5.6 Regulatory constraints

Regulatory hurdles also represent a significant barrier to the widespread adoption of 3D printing in bridge construction. Building codes and standards that govern infrastructure projects are based on traditional construction methods, and many jurisdictions lack the framework to accommodate the unique aspects of 3D printing. Updating these regulations to support innovative construction technologies while ensuring safety and quality is a complex process that requires collaboration between industry stakeholders, regulatory bodies, and researchers.

5.7 High initial costs

High initial costs are a major obstacle to integrating 3D printing technology into bridge building. The transition from traditional construction methods to innovative 3D printing techniques necessitates substantial investment in specialized equipment. This equipment, often cutting-edge and complex, comes with a hefty price tag, not only due to the technology itself but also because of the need for ongoing maintenance and updates to stay current with advancements in the field.

Additionally, the shift towards 3D printing in construction requires a workforce that is skilled in operating this new technology. Training personnel to proficiently use 3D printing equipment and understand the nuances of the process involves further financial outlays, encompassing both the direct costs of training programs and the indirect costs associated with the time employees spend away from their regular duties.

For many construction firms, particularly those that have long relied on conventional construction methods, the prospect of such significant initial expenses can be daunting. While the long-term benefits of 3D printing, such as increased efficiency, reduced waste, and the ability to create more complex structures, may indeed justify these upfront costs, realizing these benefits requires a vision that spans beyond immediate financial considerations. The transition to 3D printing in bridge construction demands not only a substantial financial commitment but also a strategic approach to financial planning and resource allocation. Firms must weigh the potential returns against the initial investments and consider the broader implications of integrating 3D printing technology into their operations, making this one of the key challenges to its broader adoption in the industry.

5.8 Lack of required skills for adopting 3D printing

The lack of required skills presents a significant barrier to adopting 3D printing in bridge construction, primarily due to the construction industry's traditionally conservative nature. This industry is characterized by well-established practices that are deeply ingrained in its operations, making the shift towards a radically different technology like 3D printing a considerable challenge. It demands a fundamental change in mindset across all levels of the organization, from the workforce executing tasks on the ground to the management teams and stakeholders overseeing projects. Bridging this divide requires thorough education and training initiatives to ensure that professionals are adequately prepared with the essential skills and knowledge for successfully applying 3D printing methods in bridge construction efforts.

Another layer of complexity is introduced by the specific demands of 3D concrete printing, which necessitates workers skilled in seamlessly integrating robotic technology with traditional civil engineering tasks. The expertise required extends to a deep understanding of the printing parameters and the thixotropic properties of the materials used, which are critical factors affecting the quality of the construction and the architectural design's feasibility [18]. As the 3D printing industry is expected to experience significant growth, professionals accustomed to conventional construction methods must adapt to these new demands. This adaptation may involve re-training existing workers or transitioning them to different roles within the industry [18]. Given these challenges, there is a pressing need for further research to comprehensively understand and address the skill gaps, ensuring that the workforce is prepared to embrace 3D printing technologies in road construction. This preparation will be crucial for overcoming the hurdles of integrating 3D printing into established construction practices and fully realizing its potential benefits.

5.9 Uncertainty in environmental advantages

The environmental and sustainability implications of applying 3D printing in highway bridge construction are complex and multifaceted, warranting a thorough examination. This examination should not only focus on the immediate benefits and challenges but also consider the broader impacts of such technological adoption on the environment and society's sustainability goals.

Mitigating the environmental effects of using 3D printing for bridge construction is essential. This process requires a thorough analysis of the materials used in 3D printing, the technology itself, architectural designs, and adherence to existing codes and standards. Important metrics for evaluation are greenhouse gas emissions, potential for acidification and eutrophication, toxicity, and the depletion of both energy and non-energy resources. Conducting an effective assessment can be achieved through advanced methodologies such as Life Cycle Assessment (LCA) or Material Flow Analysis (MFA) [20].

The potential of 3D printing technology in road construction to contribute to environmental and sustainability goals majorly lies in its ability to optimize material usage and reduce waste, which in turn can lower the carbon footprint associated with construction projects. Additionally, the precision of 3D printing allows for the creation of structures that are only as large as necessary, further conserving materials and reducing the overall environmental impact. The ability to use a variety of materials, including recycled and locally sourced materials, can also contribute to reducing the depletion of non-renewable resources and minimizing the environmental footprint of transportation and procurement processes.

Moreover, the innovation brought about by 3D printing could lead to the development of new materials and construction processes that are more environmentally friendly and sustainable. For instance, the use of materials with lower embodied energy or the integration of features that enhance energy efficiency in the final structure could be facilitated by the customizable nature of 3D printing.

There is a scarcity of research on the environmental advantages of 3D printing (3DP) compared to conventional construction methods. This gap is due, in part, to the limited number of studies evaluating the environmental performance of large-scale 3D printed buildings, which stems from a lack of reliable data—much of which originates from industry sources without rigorous research backing. Consequently, researchers often focus on smaller components, such as printed walls [3]. Other challenges include the variability of factors and scenarios involved in 3D printing, such as the equipment used for printing and pumping, the mixture's type and composition, and the geometry of printed elements [3]. Additionally, many studies exploring the environmental impacts of 3D printed concrete (3DPC) do not employ the Life Cycle Assessment

(LCA) methodology. There is also uncertainty in defining appropriate functional units for comparison in LCA studies, primarily because these studies vary in scope, with most concentrating on the material or element level rather than entire structures [3].

5.10 Job Cuts

The adoption of 3D printing in bridge construction has the potential to streamline the construction process, making it more efficient and less labor-intensive. While these improvements are beneficial from a technological and economic perspective, they could lead to significant job cuts in the construction sector [12]. This outcome is particularly concerning in countries where a substantial portion of the population relies on construction jobs. The reduction in the number of available jobs could have profound consequences on social sustainability, leading to increased unemployment rates, reduced income security for many families, and potential social unrest. Moreover, the shift towards more automated construction processes might exacerbate inequalities, as the workforce may not have the necessary skills or opportunities to transition into new roles created by these emerging technologies. Addressing these challenges requires thoughtful consideration of the social impacts of adopting 3D printing in road construction and the development of strategies to support affected workers, ensuring that technological advancements do not come at the expense of social well-being.

6. Discussion

As the industry moves towards printing more complex structures, the demand for new printable materials that are of high quality, cost-effective, and bring added value to the user will increase. This evolution necessitates an ongoing scrutiny of materials to ensure they meet these stringent criteria. Procuring the specific materials capable of achieving these outcomes can be challenging due to the precision required in their composition, the need for consistent quality, and the scalability of production. Moreover, the development of these materials must also consider environmental sustainability and the availability of raw materials, adding another layer of complexity to the procurement process in the context of 3D printing in construction.

While 3D printing has garnered significant interest across various construction industries for its potential to revolutionize traditional practices, the path towards its broader adoption is not without obstacles. For the technology to see further improvement and integration within these sectors, it is imperative to address the existing challenges that currently hinder its widespread application.

The overall environmental advantages of 3D printing, especially with the use of 3D printed concrete (3DPC), remain unclear due to several critical factors. A major issue is the rare use of Life Cycle Assessment (LCA) in research that evaluates the environmental effects of 3DPC. LCA is essential for a detailed understanding of the environmental impact throughout every phase of a product's lifecycle, from the extraction of raw materials to manufacturing, usage, and end-of-life disposal. Additionally, there is a lack of clarity in determining functional units for comparisons in LCA studies, often due to differences in the scope of these studies. Most research focuses only on the material or component level, neglecting the evaluation of complete structures or systems. This variation in the depth of studies makes it difficult to accurately measure and compare the environmental impacts of 3D printing to traditional building techniques, thereby adding to the uncertainties about the environmental merits of 3D printing in the construction sector.

The challenge of effectively reinforcing 3D printed concrete underscores a significant barrier to the wider adoption of 3D printing technology in concrete bridge construction. Addressing this issue requires innovative solutions that can seamlessly integrate reinforcement without compromising the efficiency, structural integrity, or design freedom offered by 3D printing, thereby ensuring that printed bridges are durable, safe, and cost-effective.

To mitigate job cuts resulting from the adoption of 3D printing in construction, it is imperative for the traditional construction sector to evolve alongside this new technology and find ways to remain relevant. This evolution involves reimagining roles and creating new opportunities within the industry that capitalize on the unique advantages of 3D printing.

By integrating traditional construction expertise with innovative 3D printing applications, construction professionals can enhance the value they bring to projects through improved design, efficiency, and sustainability. Education and skill development will play a crucial role in this transition, enabling the existing workforce to adapt to new methodologies and technologies. Embracing a collaborative approach between traditional construction practices and 3D printing can lead to a synergistic relationship, where each complements the other, ultimately minimizing job losses and driving the industry forward in a manner that leverages the best of both worlds.

To effectively address the challenges of adopting 3D printing in roadway construction, it is crucial to develop materials specifically designed for the Material Deposition Method (MDM) that can be extruded continuously and layered without causing deformation in the underlying layers. This requires a careful balance of viscosity and setting time to ensure that the freshly deposited layers can support the weight of subsequent layers without compromising structural integrity.

Performing finite element analysis is essential to simulate both the structural loads and the reactive properties of the materials used, particularly when working with cementitious, viscous materials. Such analyses help in understanding how these materials behave under various loading conditions, thereby enabling the design of architectural components that are not only feasible to print but also capable of meeting the rigorous demands of bridge construction. Tailoring materials to meet these specifications will play a pivotal role in overcoming the technical hurdles associated with 3D printing in the construction of durable and reliable bridges.

The adoption of 3D printing technology in bridge construction offers a promising avenue for overcoming the industry's long-standing productivity challenges. By leveraging the automation, standardization, and flexibility of 3D printing, the construction sector can achieve significant gains in efficiency, labor productivity, and overall project outcomes, marking a significant step forward in the industry's evolution.

The evolution towards additive manufacturing necessitates a paradigm shift in how projects are conceptualized, designed, and executed. There will be an increased demand for engineers who are not only proficient in traditional construction practices but also adept with digital technologies, understanding the intricacies of 3D printing. Furthermore, the adoption of this technology requires a more significant reliance on advanced computational analysis for design and verification processes, pushing the boundaries of conventional engineering approaches. This new era in construction also demands a reevaluation of standards for inspection and load testing, placing a stronger emphasis on ensuring the structural integrity and safety of 3D-printed projects. Altogether, these changes highlight the multifaceted impact of 3D printing on the construction sector, driving innovation while also challenging industry professionals to adapt and evolve.

For 3D printing in bridge construction to become more widely used, several modifications and improvements are necessary. Firstly, the development of standardized workflows and ensuring scalability are crucial steps to streamline the construction process and allow for broader adoption. Optimizing materials for better performance, along with the incorporation of recycled and environmentally friendly materials, can address sustainability concerns and reduce the environmental impact.

Additionally, focusing on energy savings and the use of local and renewable energy sources during the 3D printing process can further enhance the method's sustainability. Deploying fabrication machinery directly on-site can significantly improve efficiency by reducing transportation costs and construction time. Finally, the ability to fabricate ad-hoc parts without incurring extra costs allows for greater flexibility and customization in bridge construction projects. These modifications collectively could make 3D printing a more attractive and feasible option for widespread use in bridge construction, leading to innovative, efficient, and sustainable infrastructure development.

7. Conclusion

The technology of 3D printing is emerging as a groundbreaking innovation in highway and bridge construction, offering solutions to many urgent issues in infrastructure development. The opportunities it presents, from enhanced design possibilities and sustainability to increased efficiency and cost-effectiveness, are compelling. However, realizing these benefits is contingent upon overcoming significant technical, regulatory, financial, and cultural hurdles. As the technology continues to evolve, collaboration among researchers, industry professionals, policymakers, and educators will be critical to navigating these challenges and unlocking the transformative potential of 3D printing in bridge construction. The journey is complex, but the rewards—safer, more sustainable, and efficient highways and bridges—promise to be well worth the effort.

Clearly, there is potential to further embrace innovation in highway construction, with opportunities available for exploring the adoption of 3D printing techniques specifically for bridge construction and rapid highway pavement repairs. Despite the promising potential, the widespread adoption of 3D printing for bridges and pavements faces several challenges, including the need for large-scale printing equipment, the development of suitable materials, ensuring the structural integrity and longevity of printed surfaces, and aligning with existing construction regulations and standards. As technology advances and these challenges are addressed, it is plausible that 3D printing could become more common in the construction of bridges and pavements in the future.

Author details

Mohammadsoroush Tafazzoli*, Fatemeh Naeijian and Syeda Farwa Narjis Naqvi
Georgia Southern University, Statesboro, Georgia, USA

*Address all correspondence to: mtafazzoli@georgiasouthern.edu

References

[1] ASTM. ASTM F2792-12a Standard Terminology for Additive Manufacturing Technologies; 2010. Available from: http://www.astm.org/FULL_TEXT/F2792/HTML/F2792.html Accessed 12 March 2020

[2] Pan Y, Zhang Y, Zhang D, Song Y. 3D printing in construction: State of the art and applications. The International Journal of Advanced Manufacturing Technology. 2021;**115**(5):1329-1348

[3] Jesus M, Pessoa S, Guimarães AS, Rangel B, Alves JL. A reflection on sustainable opportunities for 3D printing in construction. In: 2nd International Conference on Construction, Energy, Environment & Sustainability. Madeira Island, Portugal; 2023

[4] Besklubova S, Skibniewski MJ, Zhang X. Factors affecting 3D printing technology adaptation in construction. Journal of Construction Engineering and Management. 2021;**147**(5):04021026

[5] van Deventer JS, White CE, Myers RJ. A roadmap for production of cement and concrete with low-CO2 emissions. Waste and Biomass Valorization. 2021;**12**:4745-4775

[6] Snellings R, Suraneni P, Skibsted J. Future and emerging supplementary cementitious materials. Cement and Concrete Research. 2023;**171**:107199

[7] Nadal A, Pavón J, Liébana O. 3D printing for construction: A procedural and material-based approach. Informes de la Construcción. 2017;**69**(546):e193

[8] El-Sayegh S, Romdhane L, Manjikian S. A critical review of 3D printing in construction: Benefits, challenges, and risks. Archives of Civil and Mechanical Engineering. 2020;**20**:1-25

[9] Tafazzoli M, Shrestha K, Dang H. Investigating barriers to the application of automation in the construction industry. In: American Society of Civil Engineers, Construction Research Congress. Des Moines, IA, USA; 2024. pp. 941-950

[10] Tafazzoli M. Construction automation and sustainable development. In: Automation and Robotics in the Architecture, Engineering, and Construction Industry. Springer; 2022. pp. 73-95

[11] Tafazzoli M, Dang H, Shrestha K. Investigating the challenges of maximizing construction automation sustainability. In: Proceedings of the 59th Annual Associated Schools of Construction. Vol. 4. Liverpool, UK; 2023. pp. 479-487

[12] Hossain MA, Zhumabekova A, Paul SC, Kim JR. A review of 3D printing in construction and its impact on the labor market. Sustainability. 2020;**12**(20):8492

[13] Self-Repairing Cities. Prototype Road-Repair Drone in Action; 2018. Available from: https://www.selfrepairingcities.com/2018/09/11/prototype-road-repair-drone-in-action/ [Accessed: April 17, 2024]

[14] 3D Printing Industry. Scientists' Automated New Asphalt 3d Printer Paves the Way for Safer Roads; 2021. Available from: https://www.3dprintingindustry.com/news/scientists-automated-new-asphalt-3d-printer-paves-the-way-for-safer-roads-186401/ [Accessed: April 17, 2024]

[15] For Construction Pros. Advanced Paving Technologies to Develop "3D Paver". Available from: https://www.forconstructionpros.com/asphalt/article/12114322/3d-asphalt-paver-to-be-developed-by-advanced-paving-technologies [Accessed: April 17, 2024]

[16] Dang H, Serne J, Tafazzoli M. Virtual reality safety training assessment in construction management and safety and health management programs. In: American Society of Civil Engineers, Computing in Civil Engineering Conference. Corvallis, OR, USA; 2023. pp. 28-35

[17] Tafazzoli M, AP L, Namian M, Al-Bayati A. Workers' fatigue in construction projects: Assessment, detection, and mitigation, a review. In: Proceedings of the 59th Annual Associated Schools. Vol. 4. Liverpool, UK; 2023. pp. 488-496

[18] Tay YWD, Panda B, Paul SC, Noor Mohamed NA, Tan MJ, Leong KF. 3D printing trends in building and construction industry: A review. Virtual and Physical Prototyping. 2017;**12**(3):261-276

[19] Barbosa F, Woetzel J, Mischke J. Reinventing Construction: A Route to Higher Productivity. New York, NY, USA: McKinsey Global Institute in Collaboration with McKinsey's Capital Projects & Infrastructure Practice; 2017

[20] Dixit MK. 3-D printing in building construction: A literature review of opportunities and challenges of reducing life cycle energy and carbon of buildings. In: IOP Conference Series: Earth and Environmental Science. Vol. 290, no. 1. IOP Publishing; Jun 2019. p. 012012

[21] Suryadi J. Examining the Labor Shortage in the Construction Industry and Possible Solutions Presented by Industry Members. San Luis Obispo, USA: California Polytechnic State University; 2018

[22] Bock T. The future of construction automation: Technological disruption and the upcoming ubiquity of robotics. Automation in Construction. 2015;**59**:113-121

[23] Perkins I, Skitmore M. Three-dimensional printing in the construction industry: A review. International Journal of Construction Management. 2015;**15**(1):1-9

[24] Niemelä M, Shi A, Shirowzhan S, Sepasgozar S, Liu C. 3D printing architectural freeform elements: Challenges and opportunities in manufacturing for industry 4.0. In: Proceedings of the 36th International Symposium on Automation and Robotics in Construction (ISARC). 2019. pp. 1298-1304

[25] Tafazzoli M, Shrestha P. Factor analysis of construction delays in the US construction industry. In: American Society of Civil Engineers, International Conference on Sustainable Infrastructure. New York, NY, USA; Oct 2017. pp. 111-122

[26] Malaeb Z, AlSakka F, Hamzeh F. 3D concrete printing: Machine design, mix proportioning, and mix comparison between different machine setups. In: 3D Concrete Printing Technology. Butterworth-Heinemann; 2019. pp. 115-136

[27] Yeon J, Kang J, Yan W. Spall damage repair using 3D printing technology. Automation in Construction. 2018;**89**:266-274

[28] Yin H, Qu M, Zhang H, Lim Y. 3D printing and buildings: A technology review and future outlook. Technology|Architecture+Design. 2018;**2**(1):94-111

[29] Ning F, Cong W, Qiu J, Wei J, Wang S. Additive manufacturing of carbon fiber reinforced thermoplastic composites using fused deposition modeling. Composites Part B: Engineering. 2015;**80**:369-378

[30] Moynihan MC, Allwood JM. Utilization of structural steel in buildings. Proceedings of the Royal Society A: Mathematical, Physical and Engineering Sciences. 2014;**470**(2168):20140170

[31] Ashraf M, Gibson I, Rashed MG. Challenges and prospects of 3D printing in structural engineering. In: Proceedings of the 13th International Conference on Steel, Space and Composite Structures. Vol. 11. Perth, WA, Australia; Feb 2018

[32] Jin YA, He Y, Fu JZ, Gan WF, Lin ZW. Optimization of tool-path generation for material extrusion-based additive manufacturing technology. Additive Manufacturing. 2014;**1**:32-47

[33] Khoshnevis B. Automated construction by contour crafting—Related robotics and information technologies. Automation in Construction. 2004;**13**(1):5-19

[34] Buchanan C, Gardner L. Metal 3D printing in construction: A review of methods, research, applications, opportunities and challenges. Engineering Structures. 2019;**180**:332-348

Chapter 6

Performance Evaluation of Low-Volume Flexible Pavement

Shankar Sabavath, Tutta Murali Krishna and CSRK Prasad

Abstract

Low-volume rural roads (LVRRs) play a crucial role in the economic, cultural and heritage development of any nation. LVRRs are part of a built-in road system in all developed and developing countries. India has a total road network of over 6.32 million kilometers in length of roadways, making it the second largest road network in the world. The LVRRs in India constitute about 75% of all categories of roads. These roads act as a lifeline to the rural communities in terms of providing access to schools, markets, medical, recreational and commercial activities. Since many of the road agencies feel that these roads carry low traffic, a sufficient maintenance budget is not allocated to keep these roads to user satisfaction. As a result, the rate of deterioration of these pavements is higher; this can be addressed by timely maintenance. It is in this context that the performance evaluation of low-volume rural road flexible pavement and the development of a performance prediction model for in-service pavements are essential. Performance models are developed using the data collected on critical distress parameters. This study investigated the significant factors that affect pavement performance at the project and the network investigated, and models were developed using an artificial neural network and genetic algorithm for the prediction of future conditions of the network.

Keywords: low-volume rural roads, flexible pavement, performance, models, pavement evaluation

1. Introduction

Rural road connectivity, as the critical element for the economic development of the nation, has been accepted and proven worldwide. LVRRs, as used in India, are referred to as low-volume rural roads (LVRRs) around the globe. India started road development with a series of plans since its independence. However, no specific attention was given to rural roads. In the absence of a central authority to monitor design/specifications, states adopted their methods in road construction, including stage development with subjective decisions under budgetary constraints almost till the end of 2000. The Government of India, being convinced of the need for the scientific development of these roads, launched the "Prime Minister's Rural Roads Program" (Pradhan Mantri Gram Sadak Yojana) PMGSY (2000) [1], with the intent of providing all-weather roads with the full association of all State Governments. The programme started with proper network planning for a realistic assessment of needs, a substantial focus on quality with a three-tier mechanism and suitable technologies in the design, construction and maintenance of roads in rural India. The programme is constantly on the lookout for new and innovative technologies while revising and updating the codes of practice with due importance to knowledge dissemination. The objectives have also been enhanced in different stages of the programme during the past two decades. Having accomplished the majority of set goals, the programme is now on a path to Asset Management (AM) to achieve sustainable rural development.

Further, a vast road network has brought connectivity to the rural areas. Earlier, these roads were constructed as gravel or earthen roads. However, these were upgraded to the engineered structure by providing a thin bituminous surfacing layer of Open Graded Premix Carpet (OGPC) of 20 mm as per the IRC: SP: 72–2015, [2] guidelines. The methods that are used for the maintenance of these roads are based on the eye judgement, ad-hocism and experience of engineers without considering actual pavement performance data; as a result, these roads are damaged after one or two monsoons. So, it is necessary to plan regular maintenance that would minimise not only the usage of financial resources but also other resources such as equipment, workforce and materials. A study conducted by [3] pointed out that the road in deplorable condition is four to five times the cost if a pavement is regularly maintained while it is in good condition. The allocation of budget towards the maintenance of low-volume rural roads is always on the lower side when compared with the high-volume highways. Further, the tendency is to neglect those roads that have low traffic volume and a low composition of light commercial vehicles. This has led to different procedures being adopted for the maintenance of LVRRs, and therefore, allocated resources should be used wisely [4].

In view of the growing importance of low-volume rural roads, the construction, maintenance and evaluation of their performance are of paramount importance. **Figure 1** shows a view of a typical low-volume rural road in India. Various research institutes and agencies have attempted over the last 30 years to develop models to predict the performance models of high-volume highway pavements. However, very

Figure 1.
View of a typical low-volume road in India.

few studies have been carried out on LVRRs. Therefore, this chapter presents the performance models developed to describe the predicted pavement condition. Using these models, one can predict the future pavement condition and plan the maintenance and budget requirements accordingly.

2. Study area and data collection

In this study, a total of fifteen road stretches were selected in three districts of Andhra Pradesh, India, to represent different geographical and environmental conditions. After identifying the test sections according to the above-said criteria, the performance evaluation of in-service pavements was conducted for 3 years during pre-monsoon and post-monsoon. The design details and historical data on test sections were compiled from the Detailed Project Reports (DPR) of the project roads. After identifying the test sections, the road inventory (Height of embankment, width of carriageway, width of shoulders and side slopes) details of the pavement sections were measured in the first round of investigation. The pavement evaluation studies were conducted over 3 years. Pavement distress due to structural failure and functional failure in terms of rutting and roughness, respectively, were measured during the same periods. MERLIN equipment was used to collect the roughness data at regular intervals along the road. The structural evaluation was done using a Portable Falling

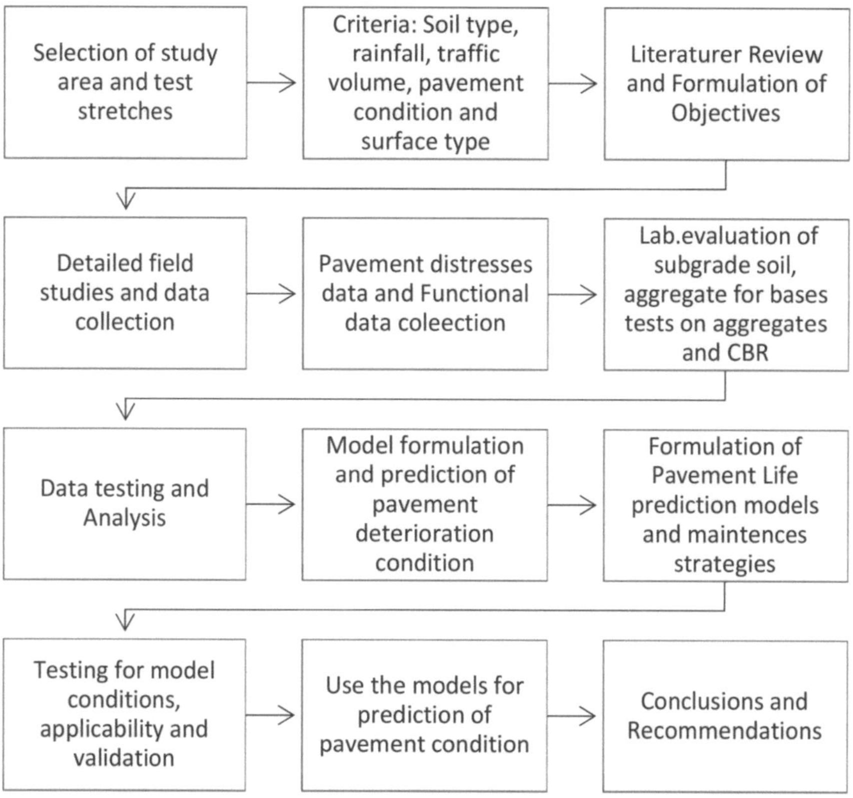

Figure 2.
Research approach.

Weight Deflectometer (PFWD) and dynamic cone penetrometer. The research approach for the development of performance models is illustrated in **Figure 2**.

3. Research approach

One of the critical tasks of this study was to investigate the significant factors that affect the pavement performance of LVRRs at project and network levels. To achieve this target, the trend between different independent parameters was plotted to examine the behaviour and interactions of different parameters considered. This has given an idea of the selection of influencing parameters for the model-building process. Taking this into account, initially, models were developed for each road level, and those models were integrated to develop project and network models for each distress. An iterative linear and non-linear regression analysis was adopted to determine the model coefficients as well as artificial neural network (ANN) and genetic algorithms techniques.

3.1 Modelling approach for pavement performance

One of the critical goals of this research was to investigate the significant factors that affect pavement performance at the project and network levels. Every likely variable that may affect pavement performance was considered initially when identifying the variables. The significant variables that are known to affect performance are pavement layer material properties, subgrade characterisation, climate, traffic, maintenance and drainage. If one or more variables adequately represent each of these parameters, then the model should contain most of the significant variables known to affect performance. A preliminary list of significant explanatory variables that affect pavement performance is prepared from previous literature.

A linear model form of the form $y = a0 + a1x1 + a2x2 + \dots + anxn$, typical non-linear power form, that is, $y = A0 + A1\ X1A2\ X2\ A3\ X3\ A4$ and $y = ex$ of exponential non-linear, etc., was used. The primary advantage of the general power form and exponential form of the nonlinear models is that they do not rely on fixed exponents, such as squares or cubic powers. An iterative linear and nonlinear regression analysis of various forms of models, viz. linear, power, exponential and logarithmic forms, was used for the development of deterioration prediction. The best model was chosen based on the significance tests, that is, the standard error (SE), t-test, F-test and coefficient of determination of the model R2 value. Basically, SE is one of the many ways to quantify the difference between an estimator and the actual value of the quantity being estimated [5].

3.1.1 Pavement cracking and pothole models

Pavements tend to crack at some point in their life under the combined effect of traffic and environment. There are a number of failure mechanisms associated with the occurrence of cracking in asphalt pavements, such as fatigue, longitudinal cracking, transverse cracking. In this study, cracking represents all types of cracking, such as fatigue cracking, longitudinal cracking, transverse cracking. The model was developed at the project level, and the network level for cracking and potholes is presented below.

$$Craking_{Project\ Level} = \left\{ \left[0.146 \times (CBR)^2 - 2.286 \times (CBR)\right] + [7.928 \times Ln(N)] + \left[0.0006 \times (LL)^2 + 0.208 \times (LL)\right] + [0.235 \times (PI)] + \left[-0.131 \times (PAGE)^2 + 1.754 \times (PAGE)\right] + 2.241 \right\} \quad (1)$$

$R^2 = 0.91$, F-test = 78.36, SE = 0.926.
(See **Table 1**)

$$Cracking_{Network\ level} = \left\{ \left[0.014 \times (CBR)^2 - 0.209 \times (CBR)\right] + [2.533 \times Ln(N)] + [-0.001 \times (LL)^2 + 0.132 \times (LL) + [0.771 \times (PI)] + \left[-0.034 \times (PAGE)^2 + 0.431 \times (PAGE)\right] + [0.067 \times (BC)] - 9.479 \right\} \quad (2)$$

$R^2 = 0.69$, F-test = 261.255, SE = 2.773.
(See **Table 2**)

$$Pothole_{Project\ level} = \left\{ \left[0.003 \times (CBR)^2 - 0.026 \times (CBR)\right] + 1.522 \times (N) + 0.096 \times (MMP) + 0.387 \times (NOP) + 0.169 \times (PAGE) + 0.090 \times (STV) - 0.067 \times (C) + \left[0.166 \times (LG)^2 - 0.364 \times (LG)\right] + 0.866 \right\} \quad (3)$$

$R^2 = 0.92$, F-test = 70.133, SE = 0.11.
(See **Table 3**)

Parameter	CBR	N	LL	PI	PAGE
t-test values	−1.14	1.36	5.4	5.3	0.29

Where CBR = California Bearing Ratio (%); N = traffic intensity (msa); LL = liquid limit of the subgrade soil (%); PI = plasticity index of the subgrade soil (%); and PAGE = pavement age (years).

Table 1.
Statistical analysis for road level.

Parameter	CBR	N	LL	PI	N	PAGE	BC
T-test values	1.80	−0.49	4.82	1.25	20.01	0.922	1.03

Table 2.
Statistical analysis for district level.

Parameter	CBR	N	MMP	NOP	PAGE	STV	C	LG
T-test values	−0.866	1.92	1.22	16.49	4.1	0.92	3.09	0.61

CBR = Subgrade California Bearing Ratio (%); N = traffic intensity (msa); MMP = mean monthly precipitation (mm); NOP = No. of potholes; PAGE = pavement age (Yrs); STV = stripping value (%); C = camber (%); LG = longitudinal gradient (%).

Table 3.
Statistical analysis for state level.

3.1.2 Development of performance models using ANNs

Artificial neural networks (ANNs) play a crucial role in pavement deterioration modelling due to their ability to capture complex relationships and patterns within large datasets. The importance of artificial neural networks in pavement deterioration modelling lies in their ability to handle complex, nonlinear relationships, adapt to changing conditions and provide accurate predictions based on data-driven insights. These attributes make ANNs valuable tools for improving our understanding of pavement performance and optimising maintenance strategies.

3.1.2.1 Pavement cracking and pothole models

Bituminous surfaces start developing cracks at some point in their service life under the combined action of traffic loading and the environment. The cracks in the surface are defects of a severe nature, which weaken the pavement structure on account of water penetration and are mainly responsible for further deterioration. Cracks once initiated progress in extent and severity and ultimately lead to potholes. The initiation of cracking is defined as the stage when a crack is observed on the pavement surface.

Cracking is a function of various parameters such as CBR, traffic intensity, liquid limit and plasticity index of the soil; mainly in this analysis, the affected parameters are taken as inputs, and the predicted parameter (cracking) is taken as output. With these input and output parameters, by using the ANN technique, the model was developed for road level, district level and state level with the same data that is used in linear regression analysis.

The neural network worksheet will be displayed with the help of the nntool function in MATLAB. In that window, the input and output parameters will be given, and that should be preceded with a training function in order to train the network with different numbers of iterations based on the quality of predictions. It is to be noted here that for three of the project levels, 90%of the data was trained for the prediction analysis, 5%of the data was validated and 5%of the data was tested. Once the whole data is tested, the overall performance of the data will be given with a target R^2 value. As the iterations in the tool function increase, the outliers in the predicted data decrease, such that there will be high accuracy in the output prediction. For the whole performance in this analysis, the prediction value will be given by the overall R^2 value with targeted outputs for cracking and pothole output, which are shown in **Figures 3** and **4**, respectively.

3.1.3 Development of performance models using genetic programming

Genetic programming is a type of evolutionary algorithm that uses principles inspired by biological evolution to evolve computer programmes to solve specific problems. In the context of pavement performance prediction, genetic programming can play a significant role for several reasons: while genetic programming can offer advantages, it is essential to note that its success depends on proper application, careful tuning and validation. It should be used in conjunction with domain knowledge and other modelling techniques to ensure accurate and reliable pavement performance predictions. In this study, a detailed analysis was carried out to develop the

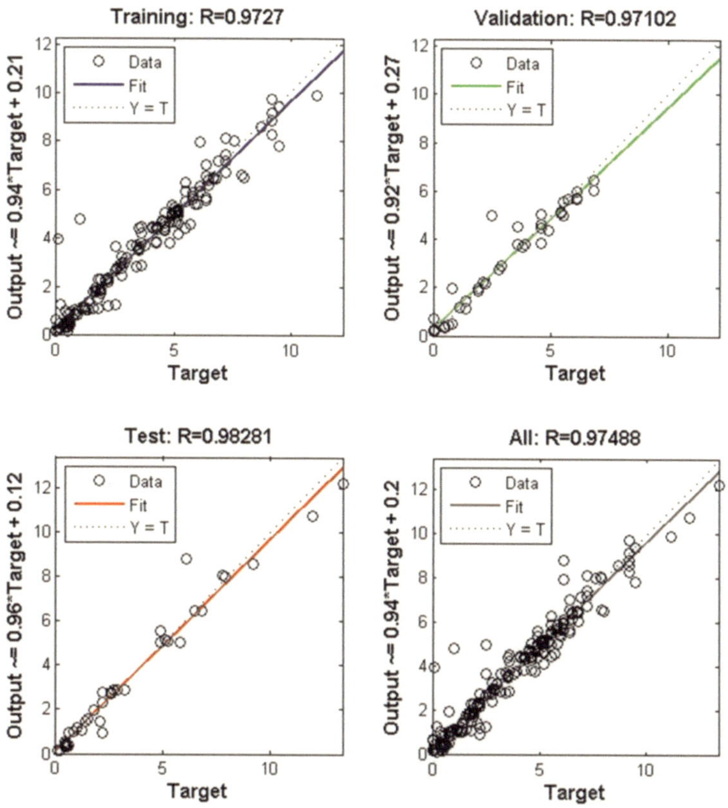

Figure 3.
Cracking output analysis.

performance models with the same data but with different techniques. With the help of genetic programming, comprehensive project-level and network-level models have been developed for cracking and pothole distresses. The simplified overall GP expression of cracking analysis for the project level and network level is presented below along with output analysis of the cracking in **Figure 5**.

3.1.3.1 Pavement cracking and pothole models

$$Cracking_{Project\ Level} = 0.02631\,LL + 9.466\,CBR^2 * PI^2 + 5.957 * CBR * PI + 2.587 * CBR * TRF * PI + 1.00 \quad (4)$$

R^2 = 0.99, SE = 0.0012, where

$$Cracking_{Network} = -832.3 * LL * TRF^2 * PI^3 - 0.03076 * LL * TRF * PI + 1.258 * CBR * TRF + 7.043 * CBR * PI - 0.163 * CBR + 0.1141 * LL + 1.004 \quad (5)$$

R2 = 0.99, SE = 0.002;

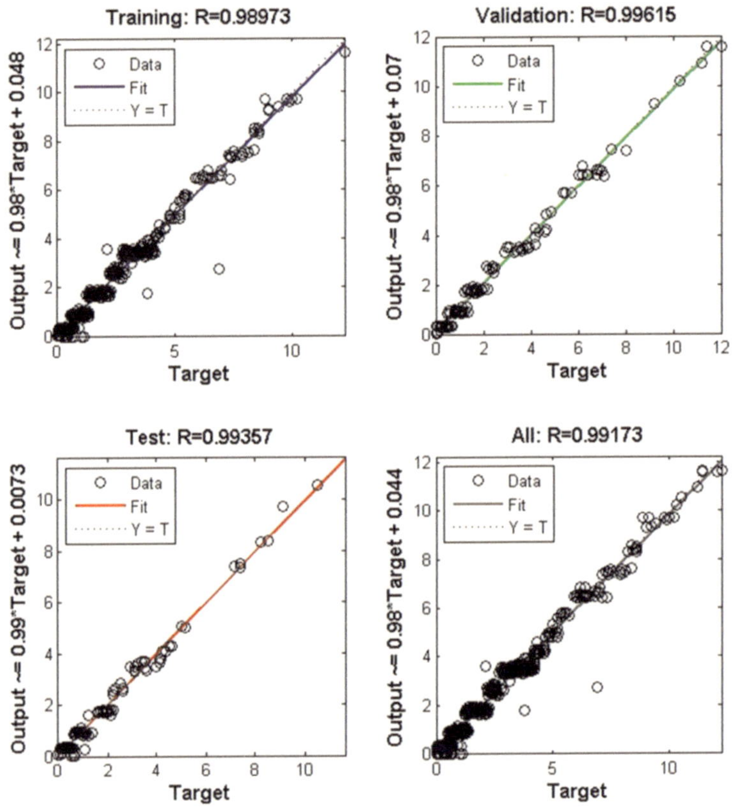

Figure 4.
Pothole output analysis.

$$Pothole_{Project\ level} = 0.1177\ CDI - 0.07674\ TRF - 0.06887\ C - 0.03837\ N + 6.702 * C * N + 0.0754 * CDI * N + 0.6862\ C^2 - 0.03837N^2 - 0.0305 * C * TRF * N - 0.0305 * TRF * CDI * N + 1.006 \quad (6)$$

R2 = 0.99, SE = 0.002

$$Pothole_{Network\ level} = 0.1122\ CDI - 0.159\ C + 0.9968 * C * TRF + 6.946 * C * N - 0.1924 * CDI * N + 1.005 \quad (7)$$

R2 = 0.99, SE = 0.002.

The output analysis of network-level data runs with the input and output variables with the Cherkassky function to predict the output with variation between actual and predicted data. The whole analysis will be displayed with different graphical representations with R-square values and a set of errors for each analysis; four graphs will be displayed, which show prediction accuracy, statistical value, a summary of the run and a set of error functions which shows overall performance shown in above **Figure 6**.

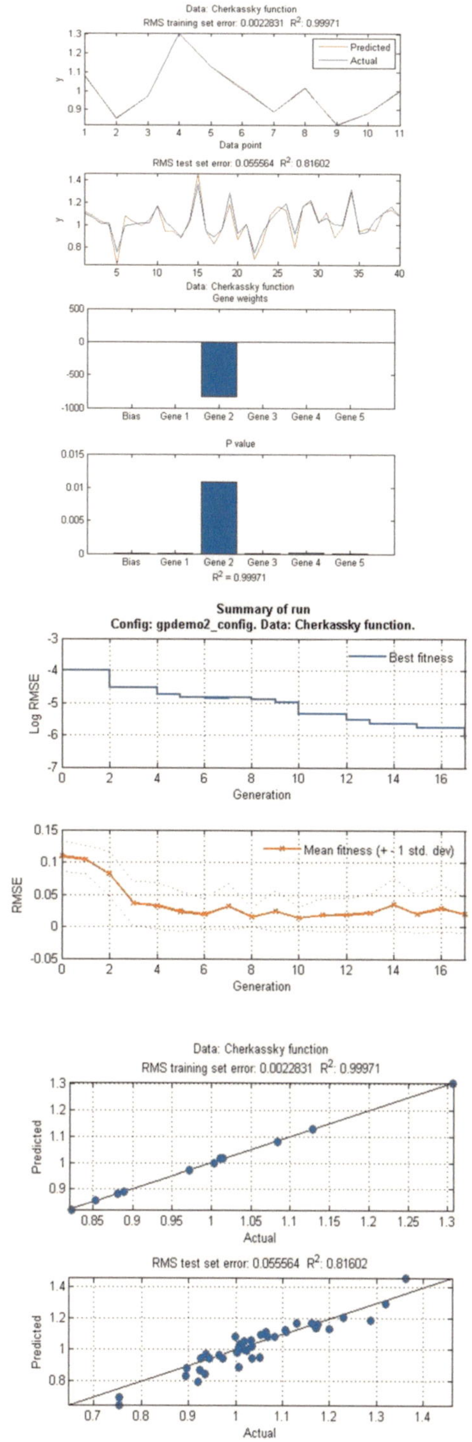

Figure 5.
Output analysis of the cracking.

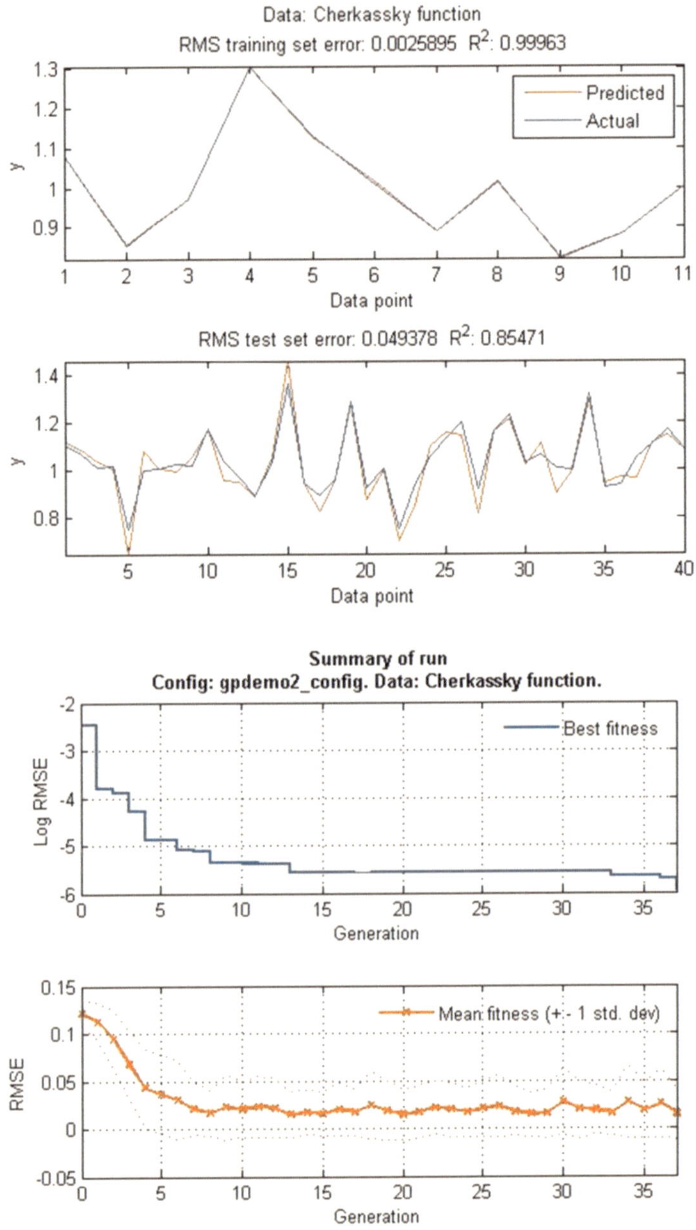

Figure 6.
Output analysis of the pothole.

4. Conclusions and way forward

Based on the results obtained from analysis work carried out in the present study, the following conclusions are drawn.

i. The performance models developed at all project and network levels showed a good prediction of pavement performance. It was also found that ANN and genetic programming give better prediction compared to linear regression at a reasonable degree of accuracy.

ii. In linear regression analysis, the overall prediction from each distress is not more accurate due to the outlier percentage being higher at all the levels in the data. The statistical values (P-value; F-value) show a degree of accuracy in simple regression analysis with less quality of fit.

iii. The ANN, the overall prediction, is better for projects and networks compared to linear regression. At the stage of testing and validation at the project level for all the distress, it was observed that the testing target prediction was more accurate than that of the validation prediction. At the network in the validation of the data, it was observed that the prediction accuracy was more than that of the testing analysis. However, in terms of overall performance, there is a good quality of fit with more precision.

iv. The overall prediction at the project level and network level in genetic programming gives much more accuracy than neural networks. The different techniques used in this study are linear regression, ANN and genetic programming. Each one has its advantages and disadvantages, but when we make a comparison among them, genetic programming gives more prediction when compared to the other techniques, which can solve both linear and nonlinear equations with basic programming coding.

As a way forward, India started systematic rural road development late but progressed notably. Along with PMGSY at the Centre-State Governments on their budget for LVRRs. However, mere asset creation does not lead to sustainable development. Nevertheless, this would be possible only when the created asset is systematically and adequately maintained, saving the created asset.

Conflict of interest

The authors declare no conflict of interest.

Author details

Shankar Sabavath*, Tutta Murali Krishna and CSRK Prasad
Transportation Division, Department of Civil Engineering, National Institute of Technology, Warangal, India

*Address all correspondence to: ss@nitw.ac.in

References

[1] IRC:SP:20-2002. Rural Roads Manual. New Delhi, India: Indian Roads Congress; 2002

[2] IRC: SP: 72. Guidelines for the Design of Flexible Pavements for Low Volume Rural Roads. New Delhi, India: Indian Roads Congress; 2015

[3] Haas R, Hudson WR, Zaniewski J. Modern Pavement Management. Florida: Krieger Publishing Company; 1994

[4] Oglesby CH. Dilemmas in the administration, planning, design, construction, and maintenance of low-volume roads. In: Low-Vol. Roads: Spec. Rep. 160. Washington, DC: Transportation Research Board; 1985

[5] Draper NR, Smith H. Applied Regression Analysis. 2nd ed. New York: John Wiley & Sons; 1981

Chapter 7

Risk Matrixes as Environmental Management Tools

Alice Elizabeth González, Martín Paz Urban, Martín Goyeneche and Lady Carolina Ramírez

Abstract

Within the framework of a broader project, the vulnerability to climate change of roadworks of art (bridges and culverts) was analyzed. A hydrological study was done; a set of preselected bridges and culverts was characterized. A methodology was proposed to work with risk matrices in order to determine an investment prioritization criterion from this point of view. This methodology was applied to the preselected set of bridges and culverts intended to have higher vulnerability to climate change (because of their years in service, their design, current scour depth, and so on). The proposed methodology was implemented in detail in two specific cases and turned out to be robust and easy to apply.

Keywords: risk matrices, climate change, vulnerability, environmental management, bridges and culverts

1. Introduction

1.1 Risk and risk assessment

Risk can be defined as the probable damage that a disaster can cause according to certain conditions of threat and vulnerability (Ferradas in [1]). Risk implies the existence of future or hypothetical damages, the occurrence of which depends on certain conditions that can be assumed to be human or environmental causes (Hurtado et al. in [1]). Some disasters are caused by environmental phenomena that cannot be neutralized, although in some cases they can be partially prevented.

Thus, risk is defined as a combination of the probability of an event occurring and its consequences, including its severity [2]. The identification and assessment of risks is carried out through the analysis of threats (natural or anthropic phenomena that may arise) and the vulnerability of potential recipients. Risks are identified and assessed through methodologies that consider the probability of occurrence of a risk event and its consequences, in order to determine the level of current risk [1].

The IPCC's approach to vulnerability and risk is now broader than the "classical" approach that used to be focused on extreme events, probability, and uncertainty. Through a comprehensive vision, Cardona et al. [3] propose that:

> *"Understanding disaster risk management as a social process allows us to change the focus of disaster response toward an understanding of disaster risk (...). This requires to know how human interactions with the natural environment lead to the creation of new hazards and how people, property, infrastructure, assets, and the environment are exposed to potentially harmful events. Furthermore, it requires an understanding of the vulnerability of people and their livelihoods, including the allocation and distribution of social and economic resources that can work for or against the achievement of resistance, resilience, and security."*

When referring to extreme events, vulnerability may be less critical than for the case of events with a higher probability of occurrence and lower intensity. However, for the latter, vulnerability has a strong relationship with small and recurring disasters. To promote more resilient societies with greater adaptive capacity, it is necessary to understand what risk is about and learn to manage it. Anticipating the occurrence of an adverse event, being able to prevent it or capitalizing on the consequences of an event to reduce the impact of future events, may contribute to reducing risk and, consequently, to better allocate resources. Reducing the exposure and vulnerability of a certain community to a type of event might increase the risk for another community in the area; this is why it is necessary to assume the need of a plan with a systemic approach [3]. Risk management aims to reduce the probability of occurrence of the adverse impacts of an event. Risk can be represented in several ways, but here it will be seen as a function of the danger or threat and the vulnerability or susceptibility of the recipient to being overwhelmed by that danger. Therefore, correctly determining the factors involved will help create a clear and representative risk matrix.

The existence of a risk will imply the presence of a disruptive agent or hazard, which is likely to cause harm. These adverse effects will be of different nature and intensity depending on the level of vulnerability of the infrastructure that is affected [4].

According to [5], risk analysis should always be a participatory process developed with the community, with or without information generation or explicit engineering participation. When using engineering analysis, two components must be quantified: the probability of the risk and the severity of its consequences. The probability of the risk is the probability of failure of the system or one of its elements. The consequences and their severity can be obtained by applying typical engineering tools, which classify the risk according to the expected consequences and whether or not they are acceptable for the cultural thresholds of the place [6].

Vulnerability and risk assessments are key strategic activities that inform both disaster risk management and climate change adaptation. They require the use of reliable methodologies that allow adequate estimation and quantification of the potential losses and consequences of a risk event for human systems, in a given exposure time. At the local level, there is usually a lack of information to carry out a detailed analysis. However, it is desirable that these evaluations can be carried out locally to achieve the best possible result for a given community [7]. Therefore, it is common to have to resort to criteria to simplify or aggregate the available information since either basic quantitative information is missing or the information available does not have the level of detail that would be useful.

The most common way to represent the probabilistic nature of hydrometeorological phenomena is in terms of a return period Tr, which is the period of time that, on average, elapses between the occurrence of phenomena of a certain intensity [8]. The aim is to model using different return periods and obtain a range of responses to different scenarios. Therefore, the dangers have to be identified, enumerated, and determined, maintaining short- and long-term perspectives, taking into account the duration and intensity of the events, and performing an analysis of the suitability of the territory, that is, prepare a diagnosis of the location area and the current condition of each bridge. The level of detail achieved by this analysis will depend mostly on the availability of the data. On the other hand, the considered characteristics of the hydrometeorological events of interest must be carefully chosen, in order to obtain a representative set of factors to evaluate.

In European countries, there are systematic studies on the risks involved in major floods, focusing on their multisectoral consequences [9, 10]. Risk indicators include the occurrence of heavy rainfall, such as a history of flooding that can generate adverse impacts on the socioeconomic system and ecosystem services. These include flash floods, the frequency of which is expected to increase, especially in summer. Directive 2007/60/EC, the current version of Directive 2000/60/EC on operational aspects of the water policy of the European Union, is currently in force [9]. Among other provisions, the current directive suggests the construction of flood risk maps and the development of flood risk management plans, with a focus on prevention, protection, and training of people. The abovementioned directive requests flood extent maps for a scenario of low probability of occurrence (the return period is not established, but some countries have used 300 years and others, such as Sweden, have used the flood of Tr = 10,000 years) and another with medium probability (Tr ≥ 100 years). The expected flood limits should be presented on a chart, together with roads, buildings, water bodies, and so on. For flood management, it may be also useful to map the water depth, the flow velocity, or its danger (as a combination of the return period and the severity of the event), among other information [10].

Article 4 of the directive indicates the following minimum contents for the preliminary flood risk assessment; regarding the content of flood danger maps and flood risk maps, their content is also established in the directive. For threat maps, three scenarios with different probabilities of occurrence must be considered [9]:

a. *"low probability of flooding or extreme event scenario;*

b. *average probability of flooding (return period ≥ 100 years);*

c. *high probability of flooding, when applicable."*

For each of them, the extent of the flooding, the water depth, and, where appropriate, the corresponding flow velocity must be indicated. For flood risk maps, the approximate number of inhabitants that may be affected, the economic activities within potentially affected areas, and the identification of facilities that may cause pollution problems during a flood must also be presented [9].

The flood emergency management system (FEMS) is also operational in the European Union [11]. The European Commission, through its EU STAR – FLOOD1 project, proposes a new way to assess and monitor the effectiveness of FEMS, applying seven indicators that must be evaluated at the national level: planning (if there is an operational plan in case of emergency due to flooding); institutional learning (e.g., post-event reviews and consultations, opportunities for knowledge exchange between responding agencies); emergency situation drills; joint work; community training; provision of resources; and recovery-based activities. A scale of one to five is used for each criterion. As both the frequency and severity of floods are expected to increase due to climate change, societies must prepare to cope. Therefore, the evaluation was implemented in five countries (France, Netherlands, Poland, Sweden, and the United Kingdom, more specifically, England) [11]. As a result, important differences were evident between different countries in terms of the risk of suffering floods. For example, the consequences on infrastructure, for example, bridges, are included in the section related to risks for economic activity. The countries that have reported a greater number of vulnerable points for events with a medium probability of occurrence are Italy, Spain, Portugal, and the United Kingdom. From this primary evaluation, several opportunities and constraints are identified to improve the effectiveness of FEMS in Europe. Therefore, the assessment framework is an important factor in impulse further research in this area [12].

An interesting precedent is presented in Ref. [13]. The work presents the development of a method that uses GIS tools to evaluate the danger, vulnerability, and exposure to flood risk in an area of Salamanca. A nonstructural measure is proposed for the prevention and mitigation of the risk of extraordinary floods. The methodology used to analyze the flood risk was based on the exploration of the physical environment, and it was used in a combined and complementary way for the hydrological-hydraulic model using GIS (ArcGis 10.1). Hazards, vulnerability, and exposure were mapped, with the final objective of evaluating risk through the multiplication of the values of these three elements [13].

1.2 The Fine's matrix

Currently, many of the quantitative tools for risk analysis derive from William Fine's matrix (1971, as mentioned in Ref. [14]), which has derived in what is currently known as the "risk matrix". Risk matrices are usually applied both in occupational safety and environmental issues, business risks, and other applications, due to the ease with which it allows interventions to be hierarchized/prioritized.

Fine proposes to jointly analyze factors of the system under study (internal factors) and threats coming from the external environment (external factors), in order to identify, assess, and prioritize the potential risks that external threats can adversely affect the internal factors of the system and, consequently, cause damage. Quantitatively, prioritization arises from putting values on these factors and assigning a risk value that is a product of the probability of occurrence of the threat, the (frequency of) exposure to the threat, and the severity of the consequences [14].

Most current methods simply multiply the probability of occurrence by the severity of the consequences or impacts. One of them, widely disseminated, is the method advocated by the INSHT of Spain in its NTP 330 [15]. This is a simplified method, based on levels or intervals of occurrence and not on values themselves. The risk level (RL) is considered to be the multiplication of a probability level (PL) by a consequence level (CL). In turn, the PL value arises from the product of the deficiency level (DL) and the exposure level (EL). The value of DL connects risk factors in terms of their possible direct relationship with the occurrence of an accident. This is a concept analogous to that of vulnerability; the higher the DL, the greater the probability that the accident will occur, as long as there is a least frequent EL (not occasional or sporadic) at the same time. Thus, the PL can be very high, high, medium or low. Regarding the level of consequences, they can be classified as mild, serious, very serious, or catastrophic (in the case of accidents, they are said to be fatal). By crossing both values, the risk level and the intervention level are obtained, which can be ordered into four categories: intervention level I (critical situation that needs urgent

	Risk levels – Consequences levels.		
Probability	**Slightly Harmful**	**Harmful**	**Extremely Harmful**
Low L	Trivial risk	Tolerable risk	Moderate risk
Medium M	Tolerable risk	Moderate risk	Major risk
High H	Moderate risk	Major risk	Intolerable risk

Table 1.
Risk factor matrix for occupational accidents (adapted from [16]).

correction); intervention level II (correct and adopt control measures); intervention level III (improve if possible; it would be convenient to justify the intervention and its profitability); intervention level IV (do not intervene, unless a more precise analysis justifies it) [15].

Universidad del Litoral [16] takes up this method and makes two valuable adjustments, which are to convert the consequences to monetary costs and to define in a simpler way the scores for the evaluation of exposure and probability (thinking about accidents at work). It also simplifies the intervention categories, according to the degree of danger of each risk: high (immediate intervention to end or treat the risk); medium (short-term intervention); or low (long-term intervention or tolerable risk) [16].

Once the different risk magnitudes have been obtained, it is necessary to obtain the severity of consequences; in occupational safety, it is quantified according to the repercussions level, which refers to the number of people exposed to the considered danger. Finally, it is possible to have an order of action priorities, always starting with those that are in the high danger-high repercussion zone. Significant risks will consider those whose degree of prioritization is high and medium with a high, medium, or low repercussion, respectively. The severity level may be reduced if corrective measures are applied that reduce any of the factors, consequences, exposure, and probability, so the importance of prioritization would change. In [16], the matrix in **Table 1** is attributed to INSHT.

Two other possible scales of interest arise from [17]: a probability of occurrence scale and a scale to justify the investment.

Probability scale: almost certainly (it is the most possible result); very possible (almost possible, 50% probability); possible (it is a rare but possible coincidence); hardly possible (it is a very rare coincidence, but it has already happened); remote (extremely rare but conceivable); almost impossible (it has never happened in several years of exposure).

The scale to justify the investment to be made is proportional to the percentage of reduction of risk. It has three levels: in-doubt investment, normally justified investment, and fully justified investment [17].

2. Case study

The case study relates to an inter-institutional experience within the framework of the Road Rehabilitation and Maintenance Program 8733-UY financed by the World Bank. It was carried out between the Department of Environmental Engineering of the Institute of Fluid Mechanics and Environmental Engineering (DIA-IMFIA) of the Faculty of Engineering of the Universidad de la República (FING-UdelaR) and the Ministry of Transportation and Public Works (MTOP) of Uruguay, through its National Administrations in Planning and Logistics, Highways, and Topography [4, 7]. The objective was to satisfy certain contractual commitments related to the resilience of road infrastructure in the face of climate change, which involved the development of three technical reports associated with the evaluation of existing infrastructure in the face of hydrometeorological events. A working group was formed ad hoc, called "Road Resilience to Climate Change Group" (hereinafter, GRVCC). The work team was very successful in technical analysis, discussion, and exchange, which allowed for building and putting into practice an effective and robust methodology; solid results were obtained to achieve the pursued objectives [4, 7, 18].

The methodology to be applied was framed in the current approach of the IPCC on vulnerability and risk [3], which focuses on events of lower intensity but with a higher probability of occurrence. The Ministry's Engineers selected a set of 20 bridges expected to have greater vulnerability to intense hydrometeorological events, due to their location, age, characteristics, and history of the last 25 years [7]. This type of event has been hitting Uruguay with increasing frequency and intensity in recent years [19]. For this set of bridges, a detailed analysis of its characteristics was carried out by modeling for events with three return periods with the free-use software HEC-RAS [20] was performed, supported by the open-source GIS software QGIS. In this case, the simulations were made by using three different precedent soil moisture conditions (dry, medium, and humid) [21] and for three return periods (Tr = 50 years, Tr = 100 years, and Tr = 1000 years) [7, 18].

The direct and indirect socioeconomic costs derived from its occurrence were evaluated, with the objective of not only anticipating attention to adverse impacts but also allocating sufficient resources to monitoring and preparation tasks and strengthening the resilience of the communities. The allocation of devoted resources to these latter aspects is part of the necessary cultural change that current societies are forced to assume due to climate change [18, 19].

Risk matrices were built to present the information obtained in a compact and clear form. In the last stage of work, contingency plans were prepared for two of the cases studied [18, 19]. The background and procedures that the National Emergencies Agency (SINAE) has implemented in conjunction with the Departmental Emergencies Coordination Centers (CECOED) were taken as a starting point, taking into account both the successful practices and the opportunities for improvement that the institutions had detected from their application [22].

3. The risk matrices used in this case study

3.1 Planning the matrices' contents

A risk matrix is a very practical technique and visual tool for analyzing and communicating information regarding the possibilities that "something", and in this case, a bridge will fail due to the intrinsic characteristics that make it vulnerable to threats, that is, what makes prone to damages due to its occurrence as well as the type and magnitude of the consequences that could be expected in case that the threat actually occurred. The matrix has a double-entry table in which vulnerability factors and danger factors are interrelated to identify and prioritize the consequences on the infrastructure that may cause different magnitudes of hydrometeorological events [7].

Vulnerability factors are those characteristics of a bridge—including its immediate environment—that make it more prone to failure. They refer to the intrinsic disposition to suffer harm. Danger factors, on the contrary, depend on hydrometeorological events rather than on the bridge. The factor that is usually considered most representative of its probability of occurrence is the return period. Thus, the methodology consists on crossing these factors and assigning them a value that reflects the probability of occurrence and the severity of events consequences. The scales to quantify probability and severity can be quantitative, qualitative, or a combination of both [7, 18].

To assemble and fill out the matrix, adequate information must be available on both the infrastructure characteristics and conditions under study and the events to be considered to identify and evaluate the resulting risks in each interaction cell. The sensitivity of this tool lies primarily in the choice of probability and severity scores given to each interaction [7, 19].

According to [23], the case study is an application that can be qualified as "*Triple baseline analysis (environmental, social and economic)*" with high-level screening. Depending on the information available, the quality of the information obtained will be determined and the list of points in which improvement is needed will be known.

In this case, the probability of occurrence has been assigned in a decreasing manner to events with a return period of 100, 200, and 500 years, while the severity of the consequences was managed according to intervals defined for each type of effect. The severity can be assigned based on experience and professional criteria: flooded area and its use, including type of crop, livestock, etc., foreseeable number of evacuees, among others. Socioeconomic criteria will also be considered, such as the characteristics of the routes that would be interrupted if a road were not passable and the economic consequences depending on the type and quantity of vehicles and freights that travel on that road [7, 18].

The impact levels considered are shown in **Table 2**, and the levels of probability of occurrence are presented in **Table 3**.

By the multiplication of the probability of occurrence and the impact severity, the results should be classified into the five classes, as defined in **Table 4**.

The results end up categorized according to the color code shown in **Table 5**, depending on the risk value [7, 18].

In the case of a bridge failure, that is to say, the bridge falling down, the vulnerability was taken as the risk value, as presented in **Table 6**; the consequences

Expected impact	Definition
Severe (3)	Impacts that could greatly affect (in quantity and severity) the community, infrastructure, and environment.
Moderate (2)	Impacts that could moderately affect (in quantity and severity) the community, infrastructure, and environment.
Low (1)	Restricted impact, with little and weak impact on the community, infrastructure, and environment.

The color code is inspired by that of a traffic light.

Table 2.
Impact levels (adapted from [7]).

Probability of occurrence	Definition
High (3)	The threat has already materialized in the location and with high frequency, or there are very strong indications of its future occurrence.
Medium (2)	The threat has manifested itself with medium frequency, or although it has not manifested itself, there are important indications of its future occurrence.
Low (1)	The threat has never materialized or has done very little, and there are no or very slight signs indicating its future occurrence.

The color code is inspired by that of a traffic light.

Table 3.
Levels of probability of occurrence (adapted from [7]).

		Impact		
		Low	**Moderate**	**Severe**
Probability	High	Moderate Risk (3–4)	Major Risk (6)	Inadmissible Risk (9)
	Medium	Tolerable Risk (2)	Moderate Risk (3–4)	Major Risk (6)
	Low	Admissible Risk (1)	Tolerable Risk (2)	Moderate Risk (3–4)

The color code is inspired by that of a traffic light.

Table 4.
Risk levels (adapted from [7]).

	Probability	1 Low; Tr = 500	2 Medium; Tr = 200	3 High; Tr = 100
Severity	1 Low	1 – Admissible	2 – Tolerable	3 – Moderate
	2 Moderate	2 – Tolerable	4 – Moderate	6 – Major
	3 Severe	3 – Moderate	6 – Major	9 - Severe

The color code is inspired by that of a traffic light.

Table 5.
Color code for risk matrices (adapted from [7]).

considered were only those related to the connection between the closest cities on both sides of the bridge and the impact on freight traffic and tourism.

3.2 Vulnerability factors

The vulnerability analysis involves considering a set of factors specific of the bridge and its circumstances (location, state of conservation, etc.), which are those that make it more prone to failure. These characteristics have been grouped according to the type and severity of the failure, for which the structure is more sensitive: factors related to the flow velocity; factors related to the occurrence of scour; and other factors that could be related to possible failure of the structure (**Table 7**) [4, 7].

The characteristics of the bridges that were finally included in the risk analysis are presented in **Table 8**. They are grouped in piles, deck and foundation characteristics, concentration of water flows, cut section, bank conditions, presence of possible obstacles, localized scour in piles or abutments, sediment accumulation, and structural status of the bridge [7].

	Low	Medium	High
Number of households affected	Up to 30	Between 30 and 100	More than 100
Number of inhabitants affected	Up to 100	Up to 300	More than 300
Services affected	No	Education	Health/Education and health
Road cut	Up to 500 m	Between 500 and 5 km	More than 5 km
Freight transport	No	Seasonal	Daily (milk products, port)
Tourism	No	Seasonal	All year
Forests, grasslands, natural fields	Up to 1 km^2	From 1 to 10 km^2	From 11 to 100 km^2
Urban and urbanized areas	Up to 0,05 km^2	From 0,05 to 0,1 km^2	More than 0,1 km^2
Dry crops and stubble	Up to 0,25 km^2	From 0,25 to 1 km^2	More than 1 km^2
Rice and stubble cultivation	Up to 0,5 km^2	From 0,5 to 5 km^2	More than 5 km^2
Beaches, dunes, fixed, and semifixed dunes	Up to 0,1 km^2	From 0,1 to 0,5 km^2	More than 0,5 km^2
Extensive crops on dairy farms	Up to 0,5 km^2	From 0,5 to 1 km^2	More than 1 km^2
Quarries, sand pits, open pit mines	Up to 0,1 km^2	From 0,1 to 0,5 km^2	More than 0,5 km^2
Horticultural zone	Up to 0,05 km^2	From 0,05 to 0,2 km^2	More than 0,2 km^2
Lost cattle	Up to 50	From 51 to 500	More than 500

Table 6.
Consequences to evaluate (adapted from [7]).

Factors related to flow velocity in the bridge section	Factors directly related to scour	Other factors related to possible failure of the structure
Span	Piles shape	Thickness of the bridge deck
Spacing between piles	Type of foundation	Structural status
Orientation related to the axis of the watercourse	State of the banks	Concentration of water flows
Location section on the watercourse	Localized scour in the piles	
Presence of singular elements that may be obstacles to flow	Localized scour in abutments	
Sediment accumulation by the bridge structure		

Table 7.
Vulnerability factors (adapted from [4]).

By using these criteria, it was obtained that almost half of the bridges in the sample were considered low vulnerability to failure, while only three were categorized as very vulnerable [7].

Parameters	Categories	Interpretation	
Piles	Piles shape	Square	These are the most unfavorable options for scour.
		Pointed	
		Circular	They are the best options for scour.
		group of cylinders	
	Number of piles	It is used together with the span to calculate the spacing between piles. If the separation is less than 10 m it is considered the highest degree of vulnerability, while if it is greater than 20 m, it will be the lowest.	
	Orientation	Perpendicular	
		skew	
Deck	Thickness	Less thickness implies greater structural vulnerability	
	span		
Type of foundations	direct	The direct foundation reduces the possibilities and magnitude of scour.	
	Pilotage		
Concentration of water flows	Yes	The concentration of flows near the bridge increases its vulnerability.	
	No		
Cut section	Straight section	Location on a straight section reduces vulnerability.	
	meander		
Bank condition	Vegetation with maintenance	Vegetated options reduce vulnerability, while bare banks increase it.	
	Vegetation without maintenance		
	bare banks		
Presence of singular elements that can result in obstacles	Yes	Free flow reduces vulnerability.	
	No		
Localized scour in the piles	Not detected	The greater the scour, the greater the vulnerability to possible structural failure.	
	moderate		
	critic		
Scour in abutments	Not detected	The greater the scour, the greater the vulnerability to possible structural failure.	
	moderate		
	critic		
Sediment accumulation by the bridge structure	visible	The presence of visible sedimentation increases the speed of passage by reducing the runoff section. This increases the vulnerability of the bridge.	
	Not visible		
Structural status	Good	If the structural condition is not good, vulnerability and the possibilities of structural failure increase.	
	Regular		

Table 8.
Considered characteristics (adapted from [7]).

3.3 Threat factors and severity of consequences

The existence of risk implies, by definition, the potential presence of danger, that is, an agent that is likely to cause harm. These agents will be of different natures and entities. To evaluate the risks, some questions must be answered, such as what could happen? When could it happen? How would it happen? What is the probability of it happening? If it happens, what would be the consequences? To do this, it is necessary to propose an analysis of the threat or danger [4, 7, 18, 19].

The threat factors are independent of the infrastructure and are linked, in this case, to hydrometeorological events. Thus, to prepare a list of threat factors, it is useful to understand danger as the probability of a harmful phenomenon occurring of a certain intensity, during a certain period of time, which will depend on the physical-geographical characteristics of the place. It is important to have enough information about the recent, historical, and projected climate in the study area. When it is possible, historical infrastructure responses to climate change will be also considered (PIEVC, 2011 as referred in [5]).

In the case of floods, it is necessary to obtain reliable data on the physical characteristics of the affected area and its population (size, distribution, density, economic and cultural characteristics, education, health coverage, among others) [7]. The effects of the storms studied refer to the destruction of physical assets (damage) and the alteration of economic flows (losses and additional costs). Consequently, it is necessary to know in the most quantitative way possible the previous situation—in general through quantitative information published by official sources or documents based on them—as well as the damages, losses, and additional costs caused by these disasters, in order to be able to carry out a quantification of its costs and impacts as precise as possible. As the study aims to quantify the greatest possible number of aspects of interest, it is proposed to quantify parameters of each bridge and its immediate environment before and after the occurrence of the adverse hydrometeorological event being studied and to attempt to quantitatively assess the impact, based on the differences between both conditions. It also includes an economic quantification [7, 18, 19].

The main consequences could be that the deck height of non-submersible bridges is exceeded, that submersible bridges remain out of service for longer than considered in their design; the increase of localized erosion in piles and abutments, and that the structure fails as a consequence of one or more of the aforementioned effects. Likewise, considering the bridge integrated into its closest environment, the severity of other consequences related to floods and, consequently, the losses generated due to the dangerous event must also be taken into account [7, 18].

It is also necessary to consider the severity of the effects of losing road connectivity. The latter is related to the type of load transported on that section of the route, the proximity to ports and national borders, the towns it connects and the services that each one has, the existence of alternative connection options with other towns during traffic interruption, among others. The evaluated consequences, which are classified according to their severity, are listed in **Table 6**, with the categories considered in each case [7, 18].

Other factors that were taken into account when assessing the severity of the consequences are the greater or lesser interdependence of the localities closest to either side of the bridge (high vulnerability was considered when the population of one of them was more than triple the population of the other), the annual average daily traffic (less than 500 vehicles was considered a low level of traffic while more than 2000 was considered high), and the proximity to borders and ports (more weight was given to the proximity to the ports of Montevideo and Nueva Palmira, and in second place, the dry borders and binational bridges) [7, 18].

The threats or dangers considered were hydrometeorological events with different return periods: 100, 200, and 500 years, and the structural failure of the bridge [7].

3.4 Applications

The risk matrices corresponding to each of the considered bridges are included in [7]. Two of the main cases are also discussed in more detail in [18]. These two cases have been chosen to be presented in this chapter. The first case refers to a bridge on National Route 18, located near the entrance to the town of Vergara, in the east of the country, and the other one is the bridge over Arroyo Colla on National Route 2. The expected impact levels matrixes and the subsequent risk matrixes are presented in **Tables 9–12.**

3.4.1 Route 18: entrance bridge to Vergara town

The entrance to Vergara town is an old bridge over Arroyo Parao. Its location is sketched on a map of Uruguay in **Figure 1(a)**. The town has another entrance, the bridge on Route 91; it is also close to International Route 8, one of the most active land freight transportation routes in Uruguay.

	Road interruption during less than 72 h*	**Long-term road interruption**	**Failure**
		(Tr = 100, 200 y 500 years)	
Number of households affected in Vergara town	Less than 30	Between 30 and 100	Between 30 and 100
Number of inhabitants affected in Vergara town	Less than 100	Less than 100	Between 100 and 300
Affected services	No	No	No
Interruption Road 18	Moderate	Moderate	Severe
Interruption Route 91	Moderate	Moderate	Moderate
Interruption Road 8	Low	Low	Low
Transport of freights	Moderate	Severe	Severe
Forest, grasslands, natural fields	Moderate	Moderate	Moderate
Urban and urbanized areas	Moderate	Severe	Severe
Dry crops and stubble	Low	Low	Low
Rice and stubble culture	Moderate	Moderate	Severe
Lost cattle	Moderate	More than 3000	More than 3000

**This type of event occurs for Tr less than 100 years. The color code is inspired by that of a traffic light.*

Table 9.
Expected impact levels for Route 18 (adapted from [18]).

		Impact		
		Low	**Moderate**	**Severe**
Probability	HIGH Road interruption of low permanence	Moderate Risk (3): Affected households in Vergara/Affected inhabitants in Vergara/ Affected services/Road 8 interruption/Dry crops and stubble	Major Elevado (6): Road 18 interruption/Road 91 interruption/Freight interruption/Forest, grasslands, natural fields/ Urban and urbanized areas/Rice and stubble culture/Lost cattle	Inadmissible Risk (9): Not detected
	MEDIUM Road interruption longer than 72 h	Tolerable Risk (2): Affected inhabitants in Vergara/Affected services/Road 8 interruption/Dry crops and stubble	Moderate Risk (4): Affected households in Vergara/Road 18 interruption/Road 91 interruption/Forest, grasslands, natural fields/Rice and stubble culture	Major Risk (6): Freight transport/Urban and urbanized areas/Lost cattle
	LOW Failure/ collapse of the bridge	Admissible Risk (1): Affected services/Road 8 interruption/Dry crops and stubble	Tolerable Risk (2): Affected households in Vergara/Affected inhabitants in Vergara/ Road 91 interruption/ Forest, grasslands, natural fields	Moderate Risk (3): Road 18 interruption/Freight transport/Urban and urbanized areas/Rice and stubble culture/Lost cattle

The color code is inspired by that of a traffic light.

Table 10.
Risk matrix for Route 18 (adapted from [18]).

The damage and losses associated with the 100-year return period flood are not very different from those with 200-year and 500-year return periods. The difference focuses on a short-term or long-term interruption of the route by the flood.

The short-term interruption of the route affects 27 households and 86 inhabitants; the long-term interruption affects 38 households and 122 inhabitants. The estimated costs of these damages are approximately US$ 1.249.000 and US$ 1.740.230, respectively. The costs of evacuating people and transitional housing are US$ 18.600 and US $ 22.600 more in each case.

No educational or health services are expected to be affected.

Freight transport is expected to be affected moderately and severely, respectively; approximately, 20.500 tons cannot reach their destination on time. In the worst case, they have an estimated cost of US$ 1.320.000.

Large areas of rice culture are expected to be affected: 6,7 km^2 in the short-term flood and 10,1 km^2 in the long-term one. In both cases, only small areas of dryland crops are expected to be affected. The estimated costs are approximately US$ 1.052.000 and US$ 1.577.000, respectively.

The loss of livestock is the last major cost to be considered: more than 3100 heads of cattle or US$ 3.845.000 in the first case and more than 3530 heads of cattle or US$ 4.378.500 in the second one.

The estimated cost of managing a short-term flood is approximately US$ 7.450.000, including damage and losses; in the case of a long-term flood, it is about US $ 9.100.000.

	Road interruption during less than 72 h*	Long-term road interruption	Failure
		(Tr = 100, 200 and 500 years)	
Number of affected households in Rosario city	More than 100	More than 100	More than 100
Number of affected inhabitants in Rosario city	Variable	More than 300	More than 300
Affected services	No	No	No
Road 2 interruption	Moderate	Moderate	Severe
Road 1 interruption	Moderate	Moderate	Moderate
Road 61 interruption	Low	Low	Moderate
Road 52 interruption	Low	Low	Moderate
Freight transport	Low	Moderate	Severe
Tourism	Low	Low	Moderate
Forests, grasslands, natural fields	Moderate	Moderate	Moderate
Urban and urbanized areas	Moderate	Moderate	Moderate
Dry crops and stubble	Low	Low	Low
Beaches, dunes, fixed and semifixed dunes	No	No	No
Extensive drops on dairy farms	No	No	No
Lost cattle	Moderate	More than 400	More than 600

**This type of event occurs for Tr less than 100 years. The color code is inspired by that of a traffic light.*

Table 11.
Expected impact levels for Route 2 (adapted from [18]).

The information on **Table 9** is used to complete the cells of **Table 10**, considering the situations in **Table 5**: high probability of occurrence and severe impacts are not admissible, while low probability of occurrence and low severity of impacts are admissible; and all intermediate ratings as shown in **Table 5**.

As a result, no inadmissible risk situations are detected, but many major and moderate risk situations have been identified as well as others of tolerable and admissible risk.

3.4.2 Bridge over arroyo Colla

The bridge over Arroyo Colla was built in 1917. Its location is sketched on a map of Uruguay in **Figure 2(a)**. It is close to Rosario town, in the department of Colonia. The town is also close to other routes: International Route 1, which connects Montevideo with the city of Colonia and with Buenos Aires (Argentina) through its international passenger port; and National Routes 52 and 61, which connect Rosario with nearby productive areas (crops, livestock, dairy farms).

This case is similar to the previous one: the damage and losses associated with the 100-year return period flood are not very different from those with 200-year and 500-

		Impact		
		Low	**Moderate**	**Severe**
Probability	HIGH Road interruption of low permanence	Moderate Risk (3): Affected services/Road 61 interruption/Road 52 interruption/Freight transport/Tourism/Dry crops and stubble/ Beaches/Extensive crops on dairy farms	Major Risk (6): Affected inhabitants in Rosario city/Affected households in Rosario city/Road 1 interruption/Road 2 interruption/Grasslands/ Urban areas/Lost cattle	Inadmissible Risk (9): Not identified
	MEDIUM Road interruption longer than 72 h	Tolerable Risk (2): Affected services/Road 61 interruption/Road 52 interruption/Tourism/ Dry crops and stubble/ Beaches/Extensive crops on dairy farms	Moderate Risk (4): Affected households in Rosario city/Road 1 interruption/Road 2 interruption/Freight transport/Grasslands/ Urban areas	Major Risk (6): Affected inhabitants in Rosario city/Lost cattle
	LOW Failure/ collapse of the bridge	Admissible Risk (1): Affected service/Dry crops and stubble/ Beaches/Extensive crops on dairy farms	Tolerable Risk (2): Affected households in Rosario city/Road 1 interruption/Road 61 interruption/Road 52 interruption/Tourism/ Grasslands/Urban areas	Moderate Risk (3): Affected inhabitants in Rosario city/Road 2 interruption/Freight transport/Lost cattle

The color code is according to that of Table 5. The color code is inspired by that of a traffic light.

Table 12.
Risk matrix for Route 2 (adapted from [18]).

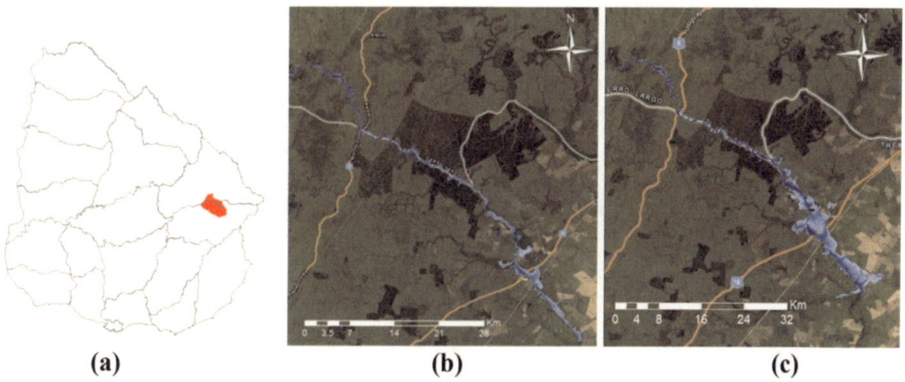

Figure 1.
(a) Location of Vergara town and its entrance bridge. (b) Affected zone in first 24 hours of flooding. (c) Affected zone for Tr = 500 years (from [7]).

year return periods. The most important thing is the permanence of the interruption of the route due to the flood.

The short-term interruption of the route affects 105 households and 336 inhabitants; the long-term interruption affects 123 households and 394 inhabitants. The estimated costs of these damages are approximately US$ 7.248.000 and US$ 8.568.000, respectively. The costs of evacuating people and transitional housing are US$ 61.000 and US$ 62.700 more in each case.

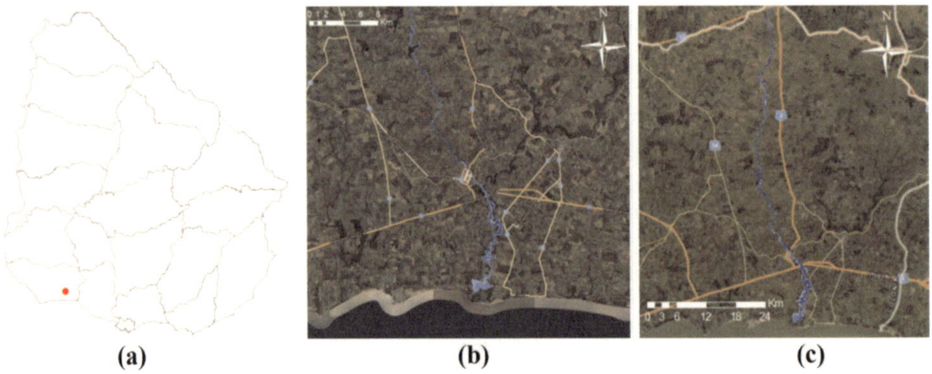

Figure 2.
(a) Location of the bridge over Arroyo Colla. (b) Affected zone in first 24 hours of flooding. (c) Affected zone for Tr = 500 years (from [7]).

Also in this case, no educational or health services are expected to be affected.

Freight transport is expected to be affected, moderately and severely respectively; approximately, 16.300 tons cannot reach their destination (e.g., Nueva Palmira International Port) on time. Given that it is an international port that there are more sensitive merchandise involved and also that the tourist connection through the port of Colonia may be affected; the losses are estimated at US$ 6.250.000.

Dryland crops areas are expected to be affected: 1,5 km^2 in the short-term flood and 1,75 km^2 in the long-term one. The estimated losses are approximately US$ 93.000 and US$ 108.000, respectively.

The loss of livestock is the last cost to be considered: 430 heads of cattle or US$ 2.304.000 in the first case and more than 650 heads of cattle or US$ 2.504.700 in the second one.

The estimated cost of managing a short-term flood is approximately US$ 16.000.000, including damage and losses; in the case of a long-term flood, it is about US$ 17.600.000. These high figures are mainly related to the large urban areas affected by the flood and the interruption of freight transportation to the Nueva Palmira International Port, the second cargo port in Uruguay.

In the same way, as the previous case, the information on **Table 11** is used to fill in the cells of **Table 12**, considering the cases in **Table 5**.

No inadmissible risk situations are detected, but many major and moderate risk situations have been identified as well as others of tolerable and admissible risk.

4. Conclusions

Successful adaptation to climate change is based on a multidimensional perspective that encompasses social, economic, environmental, and institutional aspects. Therefore, risk and vulnerability assessments intended to inform these adaptation strategies also require a multidimensional perspective.

An integrated and interdisciplinary approach makes it possible to greater certainty when addressing the complexity and dynamics of social and environmental systems and contribute to more effective risk management by the different actors involved in making decisions on risk reduction or adaptation.

Simple but effective elements, such as William Fine's risk analysis matrix adapted to the case study, may produce extremely effective results in the synthesis, assessment, and communication of information.

Acknowledgements

The authors thank the Ministry of Transportation and Public Works for financing the project through an agreement with the Julio Ricaldoni Foundation of the Faculty of Engineering, UdelaR, and for authorizing the publication of this work.

They also sincerely thank all the members of the Climate Change Road Resilience Group, the inter-institutional team that carried out this project: Arch. Rafael Ferrando, Mag. Eng. Hebenor Bermúdez, Mag. Eng. Gabriela Acosta† from MTOP; and Eng. Malena López Parard, Eng. Micaela Luzardo Rivero and Eng. Juan Ignacio Pais from DIA-IMFIA (FING-UdelaR).

Conflict of interest

The authors declare no conflict of interest.

Author details

Alice Elizabeth González[1]*, Martín Paz Urban[2], Martín Goyeneche[2] and Lady Carolina Ramírez[1]

1 Faculty of Engineering, Department of Environmental Engineering-IMFIA, University of the Republic (Udelar), Montevideo, Uruguay

2 National Administration of Highways, Ministry of Transportation and Public Works, Montevideo, Uruguay

*Address all correspondence to: elizabet@fing.edu.uy

References

[1] Renssnature & Consulting Cía. Ltda. Capítulo X Análisis de Riesgos (para Petroamazonas). Ecuador: Gruporenss; 2016. p. 2016

[2] Dirección General de Alianzas Público Privadas DGAPP. Metodología de Análisis de Riesgo. República Dominicana: DGAPP; 2020

[3] Cardona OD, van Aalst MK, Birkmann J, Fordham M, McGregor G, Perez R, et al. Determinants of risk: Exposure and vulnerability. In: Managing the Risks of Extreme Events and Disasters to Advance Climate Change Adaptation. A Special Report of Working Groups I and II of the Intergovernmental Panel on Climate Change (IPCC). Cambridge, UK, and New York, NY, USA: Cambridge University Press; 2012. pp. 65-108

[4] Grupo Resiliencia Vial al Cambio Climático. Informe: Fase I. MTOP - FJR. Montevideo, Uruguay: Grupo Resiliencia Vial al Cambio Climático; 2019. 23 p

[5] Sandink D, Lapp D. The PIEVC protocol for assessing public infrastructure vulnerability to climate change impacts: National and international application. In: CSCE 2021 Annual Conference, Canada. May 2021

[6] Engineers Canada/Ingénieurs Canadá. Public Guideline on Risk Management. Canadá: Engineers Canadá; 2020

[7] Grupo Resiliencia Vial al Cambio Climático. Producto 2: Análisis de detalle de veinte puentes. MTOP - FJR. Montevideo, Uruguay: Grupo Resiliencia Vial al Cambio Climático; 2020. 143 p

[8] Silveira L, Genta JL, Charbonnier F, Failache N, y Alonso, J. Directivas de diseño hidrológico – hidráulico de alcantarillas. Uruguay: Universidad de la República, Facultad de Ingeniería, IMFIA; Ministerio de Transporte y Obras Públicas, Dirección Nacional de Vialidad; 2000

[9] Unión Europea. Directiva 2007/60/CE del Parlamento Europeo y del Consejo, de 23 de octubre de 2007, relativa a la evaluación y gestión de los riesgos de inundación. Belgium; 2007

[10] Nixon S, Horn J, Hödl-Kreuzbauer E, ter Harmsel A, Dominique VE, Thomas D. European Overview Assessment of Member States' Reports on Preliminary Flood Risk Assessment and Identification of Areas of Potentially Significant Flood Risk. Luxembourg: Publications Office of the European Union; 2016. DOI: 10.2779/576456

[11] European Commission DG. Science for Environment Policy: Responding to Floods in Europe: New Framework Assesses Effectiveness of Flood Emergency Management Systems Environment News Alert Service, Edited by SCU. Bristol: The University of the West of England; 2017

[12] Gilissen HK, Alexander M, Matczak P, Pettersson M, Bruzzone S. A framework for evaluating the effectiveness of flood emergency management systems in Europe. Ecology and Society. 2016;**21**(4):27. DOI: 10.5751/ES-08723-210427

[13] Veleda S, Martínez-Graña A, Santos-Francés F, Sánchez-San Roman J, Criado, M. Analysis of the hazard, vulnerability, and exposure to the risk of flooding

(Alba de Yeltes, Salamanca, Spain). Applied Sciences. 2017;7(2):157. DOI: 10.3390/app7020157

[14] UNIR Revista. 2023. El método William T. Fine para el análisis de riesgos laborales - 26/12/2023. Available from: https://www.unir.net/ingenieria/revista/metodo-william-t-fine/

[15] Bellоví B, Malagón P. NTP 330: Sistema simplificado de evaluación de riesgos de accidente. España: Instituto Nacional de Seguridad e Higiene en el Trabajo, Centro Nacional de Condiciones de Trabajo, Ministerio de Trabajo y Asuntos Sociales; s.f

[16] Universidad Nacional de Litoral UNL (s.f.). Diagnóstico Situacional - Control de Riesgos. Available from: http://www.eis.unl.edu.ar/z/adjuntos/2994/Control_de_Riesgos.pdf

[17] Qué es el método de evaluación W. Fine. Available from: https://seguridadindustrial77.blogspot.com/2017/11/metodo-de-evaluacion-w-fine.html

[18] Grupo Resiliencia Vial al Cambio Climático. Producto 3: Planes de Contingencia. MTOP - FJR. Montevideo, Uruguay: Grupo Resiliencia Vial al Cambio Climático; 2020. 67 p

[19] González-Fernández AE, Paz-Urban M, Goyeneche M. Análisis de vulnerabilidad de obras viales en Uruguay ante el cambio climático. Quivera Revista de Estudios Territoriales. 2022;**24**(2):5-27. DOI: 10.36677/qret.v24i2.19349

[20] US Army Corps of Engineers, Hydrologic Engineering Center. HEC-RAS River Analysis System. User's Manual, Version 5.0. USA: US Army Corps of Engineers, Hydrologic Engineering Center; 2016

[21] Papaioannou, G., Loukas, A. y Vasiliades, L. (2019). Flood risk management methodology for lakes and adjacent areas: The Lake Pamvotida paradigm. Proceedings, 7 (21), doi: 10.3390$/$ECWS-3-05825

[22] SINAE. Informes de consultorías Generación de conocimientos en Gestión Integral del Riesgo. Guía. 2014;**6**:211-226

[23] PIEVC Global Partnership. PIEVC© Family of Resources Catalogue. A Guide for Selecting Climate Risk Assessment Methods, Data, and Supporting Materials. Canadá: Public Infrastructure Engineering Vulnerability Committee PIEVC; 2021

Section 3

Traffic Control and Safety

Chapter 8

Traffic Environment Evaluation of "Loopholes" Using ETC2.0 Data and GIS Information

Yasufumi Sekine, Yuji Hayashi, Yuya Ohtsubo, Kumiko Hamaya and Toshio Yamamoto

Abstract

During commuting hours, the number of vehicles on main line roads increases rapidly. As a result, traffic jams occur. When a traffic jam occurs, some vehicles will flow into nearby residential areas roads that can be used as a "loophole." In this study, the authors investigated multiple roads known as "loopholes" and evaluated their traffic environments using ETC2.0 data and Geographic Information System (GIS) information. It was also clarified that "loophole" roads have characteristics such as "high number of traffic volume" and "high average velocity." These are characteristics that have a negative impact on pedestrian safety in "loopholes," which have many sections without sidewalks. Therefore, it is important to take countermeasures to improve pedestrian safety on "loophole" roads.

Keywords: transportation engineering, road environment, risk assessment, human error, accident analysis

1. Introduction

Many vehicles pass on the main line road. As a result, traffic jams occur during times when the number of vehicles passing through the area increases rapidly, such as during morning and evening commuting hours. When this congestion occurs, some vehicles flow into surrounding roads that serve as "loopholes." A "loophole" road is a narrow road where the sidewalk and roadway are not completely separated or a road that is extremely narrow even if there is a sidewalk. For this reason, as the number of cars passing through a "loophole" road increases, its safety decreases. However, it is not clear whether a vehicle that enters a loophole from a main line road will return to the main line road (in Ref. [1]). Therefore, it is not clear whether the number of vehicles that enter a loophole and return to the main road increases or decreases in conjunction with the length of the traffic jam (in Ref. [2]). Furthermore, the velocity of the vehicle traveling on the "loophole" is also unknown. In this study, the authors investigated multiple roads that are considered "loopholes." Using a questionnaire survey of students at elementary schools along the road, ETC2.0 data, and GIS information, the authors analyzed the destinations, velocities, and frequency of sudden braking of vehicles passing through the loopholes. Based on these, the authors evaluated the traffic environment.

2. Investigation area and analysis method in this study

Figure 1 shows the “loopholes” and surrounding areas that were investigated in this study. The area shown in **Figure 1** includes the traffic jam section of National Highway No. 2 in Fukuyama City, Hiroshima Prefecture, and multiple “loopholes” (in Ref. [1–3]). Section (B)-(I)-(A) in **Figure 1** is the section where congestion occurs on National Highway Route 2. Then, the section (E)-(F), (L)-(M), (L)-(J), and the section (F)-(N), (M)-(O), (J)-(H)-(O) is a road that serves as “loopholes” when National Highway No. 2 is traffic jam (in Ref. [4]). In this study, the authors evaluated the traffic environment in areas that include these areas.

Regarding the above-mentioned investigation area, analysis of ETC2.0 data, questionnaire survey of elementary school students along the roadside in the investigation area (section (M)-(O)), accident occurrence situation, and destination survey of passing vehicles. (This is an investigation called a license plate investigation. This investigation uses cameras installed on the road to photograph the license plates of passing vehicles. Then, the number of vehicles passing by with the same license plate is observed at the entrance and exit of the observation target section.) were analyzed. (Using ETC2.0 data, the authors analyzed the daily traffic volume, the velocity of passing vehicles, and the frequency of sudden braking on the roads in the investigation area. In the license plate survey, the authors investigated the number of vehicles passing through section (M)-(O)).

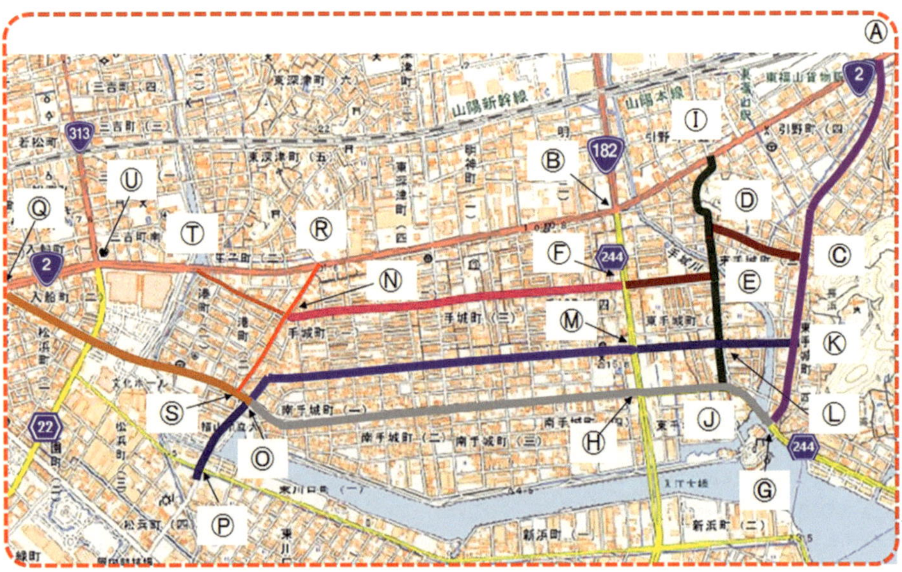

Figure 1.
Areas containing loopholes investigated in this study.

3. Traffic environment in the investigation area

3.1 Analysis of ETC2.0 data

ETC2.0 data is data stored through communication between a vehicle equipped with an ETC2.0 on-board device and communication spots installed on the road (This data is held by the Ministry of Land, Infrastructure, Transport, and Tourism of Japan). From the information recorded in the ETC2.0 data, it is possible to obtain various information such as the driving history and behavior history of the vehicle, as well as the route of the vehicle. For this reason, ETC2.0 data has been widely used for analysis of traffic conditions, traffic jams, and safety research on residential areal roads (in Ref. [5–10]). In this study, by using ETC2.0 data, the authors analyzed the daily traffic volume, the speed of passing vehicles, and the frequency of sudden braking on roads in the investigation area. (The ETC2.0 data used in this study is data for 6 months from July to December 2019).

3.1.1 Traffic conditions within the investigation area during commuting hours

Figure 2 shows the traffic conditions on roads in the investigation area during morning and evening commuting hours (Photo taken on October 20, 2020). The main line road (National Highway No. 2) at intersection (B) in **Figure 1** is congested from "east to west" during the morning and evening commuting hours (**Figure 2(a)** and **(b)**). However, the traffic jam has been resolved from "east to

(a)Intersection Ⓑ 7:03 (b)Intersection Ⓑ 17:47 (c)Intersection Ⓣ 7:00
(d)Intersection Ⓣ 16:52 (e)Between Ⓕ and Ⓝ 8:55 (f)Between Ⓡ and Ⓢ 8:55
(g)Between Ⓜ and Ⓞ 7:50 (h)Between Ⓜ and Ⓞ 17:28 (i)Between Ⓜ and Ⓞ 19:00

Figure 2.
Morning and evening commuting hours on roads within investigation area.

west" of the intersection (T) (**Figure 2(c)** and **(d)**). The road between intersection (F) and intersection (N) is a narrow road without a center line. Many vehicles pass through this road during morning and evening commuting hours (**Figure 2(e)**). Furthermore, the road in section (R)-(S), which intersects the road in section (F)-(N) at intersection (N), is also narrow. There are many vehicles passing through this road (**Figure 2(f)**). The road between intersection (M) and intersection (O) has one lane on each side and has a center line. Therefore, there are more vehicles passing through this road than in sections (F)-(N). However, the sidewalks on this road are narrow. Furthermore, although there are guardrails on the sidewalk side of the gutter, there are places where they are not installed on the roadway side (**Figure 2(g)**, **(h)**, and **(i)**).

3.1.2 Daily traffic volume on roads in the investigation area

The authors quantitatively analyzed the road traffic environment in the investigation area shown in **Figure 2** by using ETC2.0 data. **Figure 3** shows the daily traffic volume on roads in the investigation area. Overall, traffic volume on main line roads is high, and traffic volume on residential areal roads is low. However, the sections (C)-(D)-(E)-(F), (I)-(D)-(E)-(L)-(M), (L)-(J), and their connections sections (F)-(N) and (M)-(O) have a high traffic volume because they are "loopholes" when the main line road (National Highway No. 2) is traffic jam. For example, section (F)-(N) is a narrow road without a center line, but there are many places where the traffic volume is "more than 500 vehicles." Section (M)-(O) is a road with one lane on each side and a center line and is somewhat wider than section (F)-(N). In section (M)-(O), there are many places where the traffic volume is "1000 vehicles or more," and there are some places where the traffic volume is "5000 vehicles or more." In this way, even if the roads that are considered "loopholes" are narrow, the traffic volume on them is much higher than on ordinary residential areal roads.

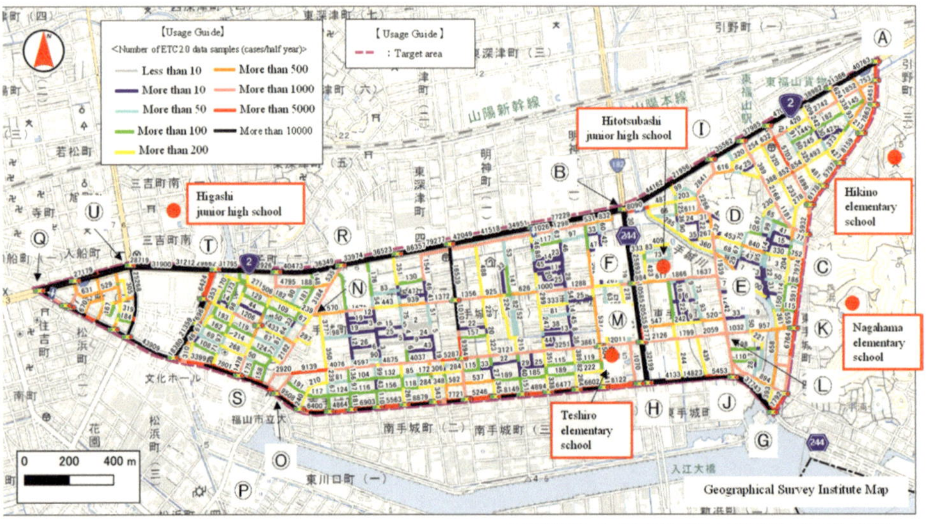

Data: ETC2.0 probe data Driving history (July 2019 to December 2019)

Figure 3.
Daily traffic volume on roads in investigation area.

3.1.3 Average velocity of passing vehicles

Figure 4 shows the distribution of average velocities of vehicles traveling on roads within the investigation area. Similar to traffic volume, the average velocity of vehicles on main line roads is "40 km/h or more" in many places. Additionally, the average velocity on residential areal roads other than loopholes is often "less than 20 km/h." However, there are many places where the velocity is "20 km/h or more" on residential areal roads that are used as "loopholes" roads. For example, in sections (E)-(F), (E)-(J), (F)-(N), and (M)-(O), there are some places where the velocity is "30 km/h or more." In particular, there are places in section (M)-(O) where the average velocity is "40 km/h or more." (This average velocity is about the same velocity as main line roads).

3.1.4 Frequency of sudden braking

Figure 5 shows the frequency of sudden braking of vehicles traveling on the roads in the investigation area. Traffic jams occur on the main line roads due to the high volume of vehicle traffic. Sudden braking occurs frequently on this main line roads. In addition, sudden braking rarely occurs on residential areal roads that are not "loopholes" during main line road traffic jam. On the other hand, sudden braking occurs more frequently on residential areal roads that serve as "loopholes." Then, on "loophole" residential areal roads, sudden braking occurs near intersections with many vehicles passing through (e.g., points (D) and (E)) or near traffic lights (e.g., near intersection (O)). The frequency of sudden braking is higher than that of other "loopholes."

3.2 Questionnaire for elementary school students in the investigation area

Section (M)-(O) has a high "daily traffic volume" and the "average velocity" of passing vehicles is also high (**Figures 3** and **4**). Furthermore, the frequency of sudden

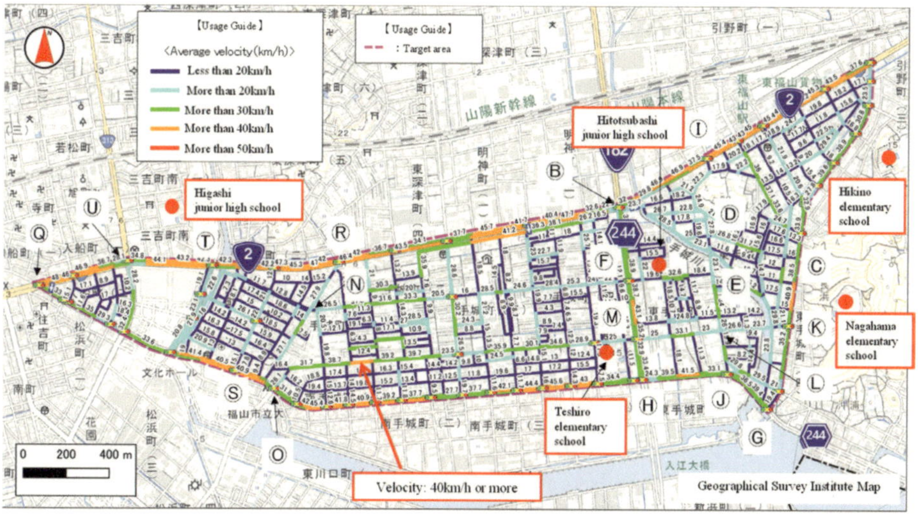

Data: ETC2.0 probe data Driving history (July 2019 to December 2019)

Figure 4.
Average velocity of passing vehicles in investigation area.

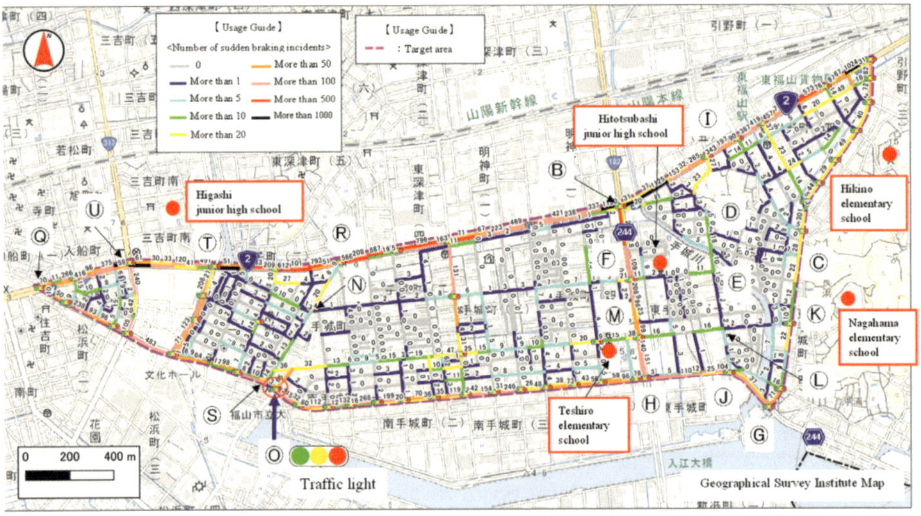

Data: ETC2.0 probe data Driving history (July 2019 to December 2019)

Figure 5.
Frequency of sudden braking in investigation area.

braking is also high around the intersection (O) (**Figure 5**). Therefore, the authors conducted a survey of students who commute to the elementary school (Teshiro Elementary School) near the intersection (M) in order to evaluate the safety around section (M)-(O). The subjects of the questionnaire were fifth graders (63 students: 30 boys and 33 girls) and sixth graders (28 students: 14 boys and 14 girls) who agreed to the questionnaire among the fifth and sixth graders at the same school. In this investigation, the authors asked elementary school students to indicate on a map the places where they had "dangerous experiences."

Figure 6 shows the results of the questionnaire. Although the road in section (M)-(O) is the road that elementary school students take on their daily commute to school, many students answered that they had "had a dangerous experience" on this section (M)-(O).

3.3 Accident occurrence status in the investigation area

Section (M)-(O) has a high "daily traffic volume" and the "average velocity" of passing vehicles is also high (**Figures 3** and **4**). Furthermore, the frequency of sudden braking is also high around the intersection (O) (**Figure 5**). The authors analyzed the occurrence of traffic accidents in the investigation area using accident data from the Geographic Information System (GIS) of the Institute for Traffic Accident Research and Data Analysis (ITARDA).

Figure 7 shows the locations of accidents that occurred within the investigation area from 2013 to 2017 on a map. In the ETC2.0 data analysis, the main line road section (B)-(R), section (H)-(O), and residential areal road section (R)-(S), sections (M)-(O), etc., there are many "vehicle-to-vehicle" accidents. In addition, there are many "pedestrian accidents" and "bicycle accidents" in section (M)-(O). Furthermore, in the results of a survey of elementary school students, many students answered that they had "had a dangerous experience" in sections (M)-(O). Therefore, the traffic environment in section (M)-(O) is considered to be a "less safe" environment.

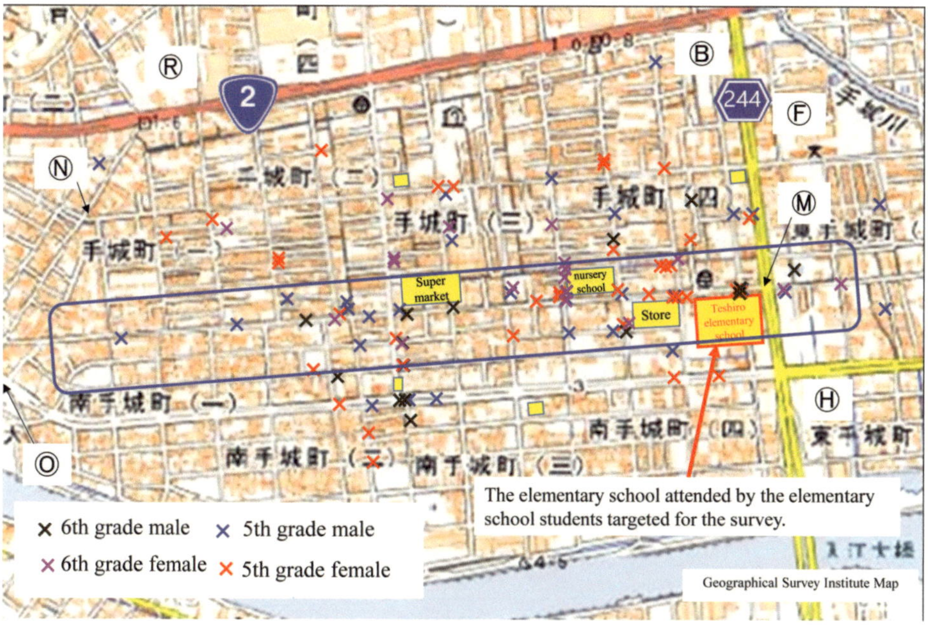

Figure 6.
Results of a questionnaire of elementary school students in investigation area.

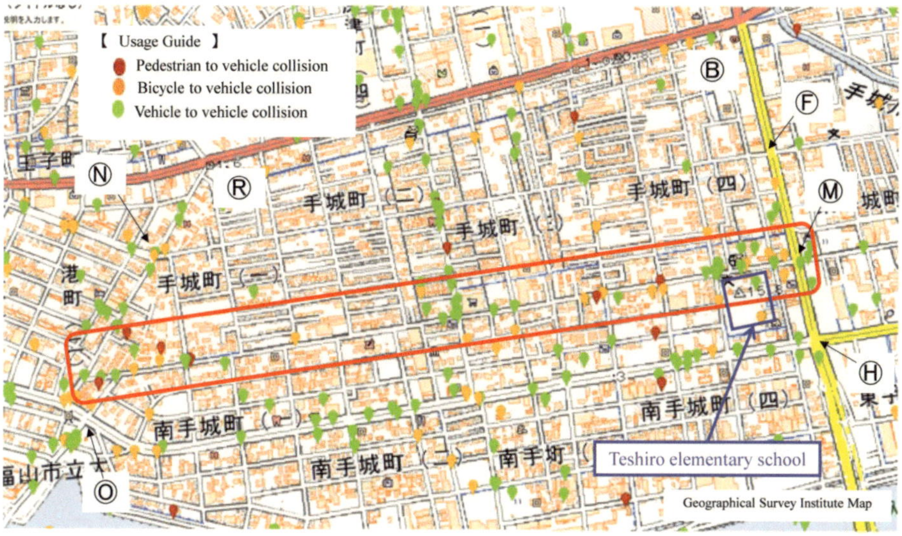

Figure 7.
Results of accidents that occurred in investigation area.

3.4 Destination of passing vehicles in section (M)-(O) (license plate investigation)

Based on the results of a questionnaire survey of elementary school students along the roads in the survey area and the results of an analysis of the occurrence of traffic accidents, it became clear that section (M)-(O) had a traffic environment with low safety. Therefore, the authors decided to analyze the destinations of vehicles passing through section (M)-(O). Therefore, the authors conducted an observational investigation to clarify the destinations of vehicles passing through section (M)-(O). The results of the observation investigation are shown in **Table 1** and **Figure 8**. At many times of the day, there are many vehicles going in the (Q) direction. Therefore, it is assumed that many of the vehicles passing through section (M)-(O) turn right at intersection (O) and try to return to the main line road National Highway No. 2) where the traffic congestion has cleared. Section (M)-(O) is a road with one lane on each side and a center line, making it easier to drive on than section (F)-(N), which has no center line and a narrow road width. On the other hand, section (H)-(O) is a wide road with a median strip, but there are more traffic lights installed along the road than section (M)-(O). For the above reasons, it is thought that many vehicles try to return to the main line road by passing through the section (M)-(O) and turning right at the intersection (O). (The observation investigation was conducted on Thursday, October 5, 2022).

	Ⓟ direction	Ⓠ direction
7:00–8:00	3	20
8:00–9:00	4	14
9:00–10:00	4	13
15:00–16:00	3	8
16:00–17:00	10	8
17:00–18:00	9	15
18:00–19:00	1	4
19:00–20:00	4	3

Table 1.
Result of license plate investigation, at (O).

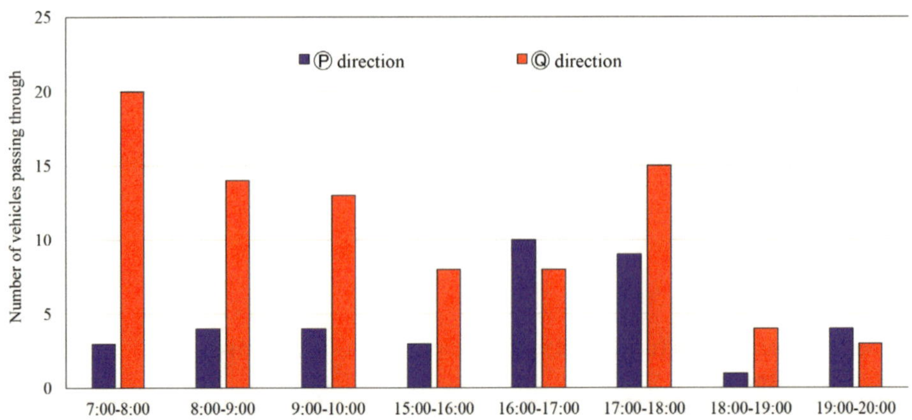

Figure 8.
Number of vehicles by destinations by time zone.

4. Discussion for traffic environment of the investigation area

The authors clarified the traffic environment of the "loopholes" that are the subject of this study by conducting the investigations and analyses described in (1) ETC2.0 data on vehicle traffic volume and average velocity, (2) Frequency of sudden braking, (3) Accident occurrence status using GIS information, (4) Questionnaire of elementary school students near the intersection (M), and (5) Destinations of vehicles traveling on road section (M)-(O).

In other words, during times when traffic jam occurs on the main line road (National Highway No. 2), the number of vehicles using the "loophole" residential areal road increases. Furthermore, if there are multiple residential areal roads parallel to the congested main line road, there will be multiple "loophole." Among the multiple "loopholes," the road in section (M)-(O) had a high volume of vehicle traffic, a high average velocity, and a high frequency of sudden braking. Furthermore, this road was one on which elementary school students answered in a questionnaire that they had "had a dangerous experience," and many accidents have occurred on this road. Among the vehicles passing on this road, many try to turn right at the intersection (O) and return to the main line road. This chapter considers the above.

Table 2, **Figures 9** and **10** show the results of an observation investigation of the traffic jam length in the (A)-(B) direction on the main line road (National Highway No. 2) at intersection (B) and the time required to pass through intersection (B). The length of traffic jams exceeds 200 m between 7:10 and 8:20 in the morning and between 17:40 and 18:40 in the evening. The required time to pass through an intersection is around 240 seconds between 7:10 and 8:20 in the morning (328 seconds at 8:10) and 300 seconds between 17:40 and 18:10 in the evening. (The observation investigation of traffic jam length was conducted on 2 days, October 20, 2000, and October 5, 2022.) Therefore, the traffic jam length and required time to pass through the intersection in **Table 2**, **Figures 9**, and **10** are 2 days. (The average value of each observation investigated is listed.) In other words, it can be assumed that there are many drivers of vehicles traveling on the main line road (National Highway No. 2) who want to avoid long traffic jams. Therefore, the number of vehicles entering the "loophole" that runs parallel to the congested main line road will increase. If there are multiple "loopholes," the following can be assumed. In other words, a road that is easy to drive on or a road that allows easy return to the main line road is selected.

Next, consider the characteristics of each of the multiple "loopholes." **Figure 11** shows roads in sections (F)-(N), roads in sections (M)-(O), and roads in sections (H)-(O). Section (F)-(N) is a narrow road, and section (M)-(O) is a road with one lane on each side. Section (H)-(O) is a wide road with three lanes on each side, and there is also a median strip. Therefore, when driving a car, it seems easier to drive on the wide road section (H)-(O). However, in the reality, there are many vehicles traveling on the road in section (M)-(O). **Figure 12** shows the distribution of traffic lights in the survey area in this study. There are "traffic lights" installed in many places on the road in section (H)-(O). In other words, it is presumed that "Even though the road between section (H) and (O) is wide and easy to drive, drivers avoid driving along this route because there are many traffic lights." In addition, in order to enter the road in section (H)-(O) from the main line road (National Highway No. 2), you will need to enter from intersections (G), (J), and (H). However, traffic jams are also occurring near intersections (G) and (H). Even if the driver understood that the road in section (H)-(O) is an easy road to drive on, this road is located in a location that is difficult to navigate from the main line road (National Highway No. 2). In

Time	Length of Retention (m)	Length of traffic jam (m)	Required time (sec)	Number of waiting at traffic lights	Time	Length of Retention (m)	Length of traffic jam (m)	Required time (sec)	Number of waiting at traffic lights	Time	Length of Retention (m)	Length of traffic jam (m)	Required time (sec)	Number of waiting at traffic lights
07:00	335	185	273.5	2	15:00	195	15	109	1	17:00	245	105	181.5	2
07:10	400	240	272	2	15:10	235	70	118.5	1	17:10	225	115	189	2
07:20	365	195	237	2	15:20	220	55	197	2	17:20	285	120	238.5	2
07:30	380	200	242.5	2	15:30	315	150	169.5	2	17:30	325	180	221.5	2
07:40	330	155	227	2	15:40	305	135	134.5	1	17:40	330	245	331.5	3
07:50	385	235	290.5	2	15:50	280	65	139.5	1	17:50	340	235	328	3
08:00	410	250	271	2	16:00	160	5	48	1	18:00	420	280	387	3
08:10	420	255	328.5	3	16:10	280	105	212.5	2	18:10	385	250	360.5	3
08:20	385	235	267.5	2	16:20	350	145	188	2	18:20	380	255	258	2
08:30	240	135	133	1	16:30	310	135	194.5	2	18:30	335	170	190	2
08:40	345	185	141.5	1	16:40	240	95	123	1	18:40	360	210	337.5	3
08:50	270	110	166.5	2	16:50	235	70	131	1	18:50	310	160	260.5	2

Table 2.
Traffic jam at intersection (B) on the main line road (National Highway No. 2). (Average of observation data on October 20, 2020, and October 5, 2022).

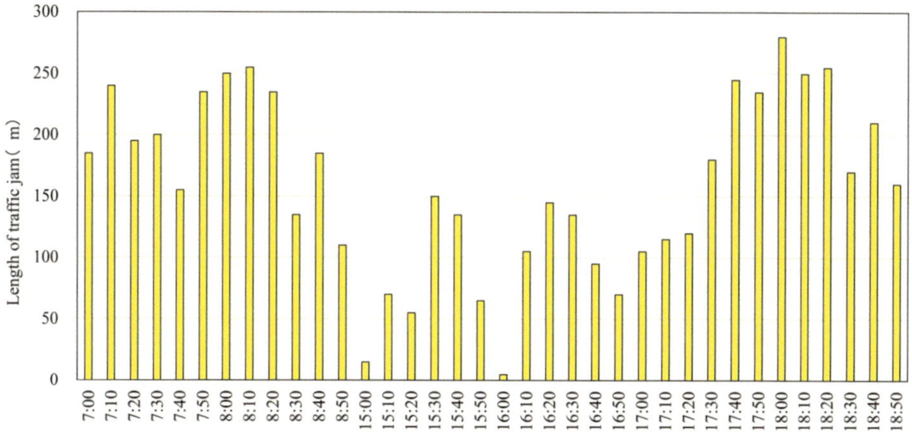

Figure 9.
Length of the traffic jam at intersection (B) on National Route No. 2.

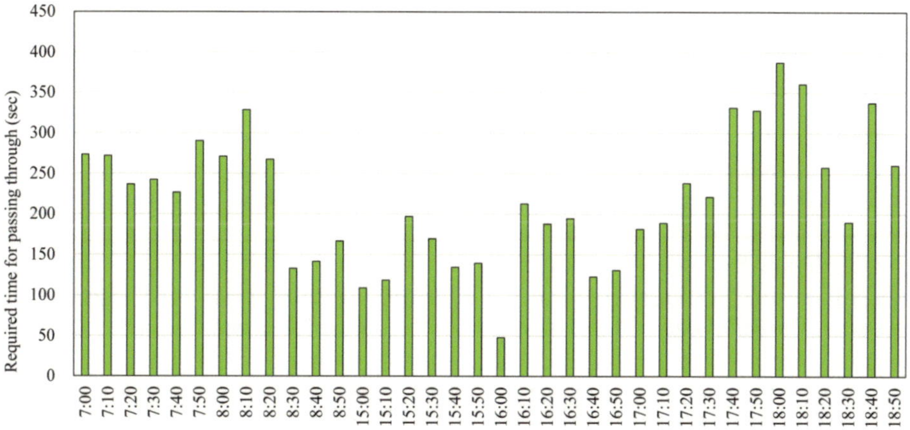

Figure 10.
Time required to passing through intersection (B) on National Route No. 2.

Figure 11.
Traffic conditions of multiple "loopholes".

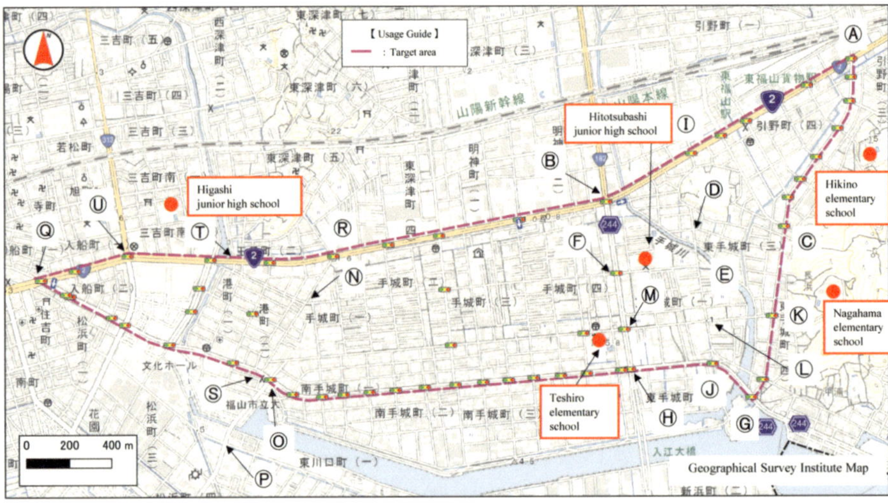

Figure 12.
Distribution status of "traffic lights" in the survey target area.

such a case, there are not many drivers who want to go to section (H)-(O). On the other hand, the road in section (F)-(N) is narrow and it is difficult to pass oncoming vehicles. Furthermore, the road leading back to the main line road (National Highway No. 2) is very narrow (Road section (R)-(S) shown in **Figure 2(f)**). Therefore, driving is difficult. The road in section (M)-(O) has a center line, so it is easier to pass oncoming vehicles than on the road in section (F)-(N). Also, there are fewer "traffic lights" than the roads in section (H)-(O). For these reasons, it is thought that many vehicles choose the road in the section (M)-(O) and try to return to the main road (National Highway No. 2)) from the intersection (O). Based on the above, the following can be considered. When multiple loopholes exist, drivers select the route to take based on conditions related to ease of driving and velocity of delivery to the destination, such as road width and number of traffic lights. However, as shown in **Figure 11(b)**, the road in section (M)-(O) has very narrow sidewalks and deep ditches. Therefore, in a questionnaire of elementary school students, many respondents said they had "had a dangerous experience" on this road. In addition to accidents between cars, there are also many pedestrian and bicycle accidents. Therefore, an increase in the number of cars on the road in section (M)-(O) is not desirable from the standpoint of pedestrian and bicycle safety. (For roads in section (M)-(O), it is also important to consider the following safety countermeasures; First, as a short-term countermeasure, there is a countermeasure to expand the space of the sidewalk by putting a lid on the gutter. Next, as a long-term countermeasure, for example, by building a road such as a bypass, vehicles that have no use in the city will be directed to another route. This will reduce traffic jam on the main road (National Highway No. 2)).

5. Conclusion

In this study, the authors investigated multiple "loopholes" roads during congestion on the main line road (National Highway No. 2). Through analysis of ETC2.0 data and traffic accident analysis using GIS information, a questionnaire investigation for elementary school students, and an investigation of the destinations of passing vehicles (license plate survey), the traffic environment of the "loophole" road was evaluated.

Analysis of ETC2.0 data revealed that "loophole" roads have the following characteristics; The daily traffic volume on residential areal roads that are "loopholes" is higher than on residential areal roads that are not "loopholes." The average velocity is also "less than 20 km/h" on many of the "non-loophole" residential areal roads, but many of the "loophole" residential areal roads are "more than 20 km/h." In particular, there are some spots on the road in section (M)-(O) where the velocity is "40 km/h or more." In addition, the frequency of sudden braking on community roads that are "loopholes" is higher than on community roads that are "non-loophole."

Among the "loophole" roads, section (M)-(O) has a high traffic volume and a high average velocity. Moreover, the frequency of sudden braking is by far the highest. The road in section (M)-(O) is the road on which many elementary school students answered in the questionnaire that they had "had dangerous experiences." Incidentally, accident analysis using GIS revealed that there are not only "vehicle to vehicle accidents" but also "pedestrian accidents" and "bicycle accidents." Additionally, many vehicles passing on this road try to turn right at the intersection (O) and head toward the main line road (National Highway No. 2) where the traffic jam has cleared.

On the other hand, when multiple loopholes exist, drivers select the route to take based on conditions such as ease of driving and speed of delivery to the destination. In the area analyzed in this study, the road in section (M)-(O) corresponds to a "loophole" that is "chosen by drivers."

However, the roads in section (M)-(O) are unsafe for pedestrians and bicycles, with very narrow sidewalks and deep gutters. Therefore, the following measures are important to ensure the safety of this road. For example, turn gutters into culverts to expand sidewalk space. Or, by constructing a bypass, vehicles that have no use within the city will be diverted to a different route.

Acknowledgements

This study is a contract research conducted in FY2018, FY2019, FY2020, and FY2020 based on the Comprehensive Collaboration and Cooperation Agreement with the Ministry of Land, Infrastructure, Transport and Tourism, Chugoku Regional Development Bureau, Fukuyama River and National Highway Office. The authors would like to express their deep gratitude to the agency for giving us the opportunity. The authors would also like to express their deep gratitude to Fukuyama City Teshiro Elementary School principal Kayoko Miyamoto and other teachers and students at the school who cooperated with the questionnaire and Ryuhei Taniguchi and other researchers Survey Research Center who carried out the survey by observing the number of passing vehicles.

Author details

Yasufumi Sekine[1*], Yuji Hayashi[2], Yuya Ohtsubo[2], Kumiko Hamaya[2]
and Toshio Yamamoto[3]

1 Fukuyama University, Hiroshima, Japan

2 Oriental Consultants Co., Ltd., Tokyo, Japan

3 Institute for Traffic Accident Research and Data Analysis (ITARDA), Tokyo, Japan

*Address all correspondence to: sekine_y@fukuyama-u.ac.jp

References

[1] Sekine Y, Yamamoto T, Hayashi Y, Ohtsubo Y. Safety of residential areal roads parallel to the main line road beyond the traffic jam intersection (evaluation of the residential areal road safety by using ETC2.0 data and GIS information). In: Of Japan Society of Mechanical Engineers 2022 Annual Conference; 11-14 September 2022; No. 22-1. J181-06. Shinjuku-ku, Tokyo, Japan: Japan Society of Mechanical Engineers (JSME); 2022

[2] Sekine Y, Yamamoto T, Hayashi Y, Ohtsubo Y. Analysis of vehicle trips and accidents on residential areal roads parallel to the main road (traffic route analysis using ETC2.0 data and accident analysis using GIS information). Transactions of the JSME (in Japanese). 2022;**88**(912):21-00350. DOI: 10.1299/transjsme.21-00350. Available from: https://www.jstage.jst.go.jp/browse/transjsme/88/912/_contents/-char/ja

[3] Sekine Y, Yamamoto T, Hayashi Y, Ohtsubo Y. An analysis of the relationship between traffic congestion on main roads and the traffic environment in surrounding areas (an evaluation of residential road safety from analysis of traffic accident and ETC 2.0). Transactions of the JSME (in Japanese). 2021;**87**(898):20-00392. DOI: 10.1299/transjsme.20-00392. Available from: https://www.jstage.jst.go.jp/browse/transjsme/87/898/_contents/-char/ja

[4] Sekine Y, Hayashi Y, Hamatani K. An analysis of the relationship between traffic jam on main line roads and occurrence of detour traffic (evaluation of traffic jam conditions and detour route estimation using ETC2.0 data). In: Of Japan Society of Mechanical Engineers 2023 Annual Conference; 3-6 September 2023; No. 23-1. J181-02. Shinjuku-ku, Tokyo, Japan: Japan Society of Mechanical Engineers (JSME); 2023

[5] Sawada H, Ohkuni M, Murakami T, Matsumoto T. A study on generation of real-time information about traffic congestion on expressway by using ETC2.0 probe data. In: Proceedings of 37th Annual Congress (2017). Japan Society of Traffic Engineers Journal of Traffic Engineering. Chiyoda, Tokyo, Japan: Japan Society of Traffic Engineers (JSTE). pp. 417-420

[6] Kanoshima H, Sizuki K, Nonaka Y, Makino H. Application of ETC2.0 probe data to bottleneck analysis on expressways. In: Proceedings of 35th Annual Congress (2015). Japan Society of Traffic Engineers Journal of Traffic Engineering. Chiyoda, Tokyo, Japan: Japan Society of Traffic Engineers (JSTE). pp. 215-221

[7] Nagahiro Y, Nishioka S, Okamoto H. Improving accuracy of travel time information by neural network using ETC2.0 data. Japan Society of Traffic Engineers Journal of Traffic Engineering, Special Issue B (Practical Papers). 2019;**5**(2):42-48

[8] Narishima S, Kasai M, Xing J, Goto H, Tsuji M. Analysis of vehicle speeds on two-way two-lane expressways with ETC2.0 data. In: Proceedings of 36th Annual Congress (2016). Japan Society of Traffic Engineers Journal of Traffic Engineering. Chiyoda, Tokyo, Japan: Japan Society of Traffic Engineers (JSTE). pp. 1-6

[9] Odaka S, Yoshii M, Kanbe N. An estimation method of traffic accident risk on residential road using the ETC2.0 probe. Japan Society of Traffic Engineers Journal of Traffic Engineering, Special Issue A. 2018;**4**(1):246-251

[10] Tanaka H, Teraoku J, Sano K, Sonobe K, Morimoto A. A study on traffic accident analysis of community roads through the use of ETC 2.0 probe data. In: Proceedings of 38th Annual Congress (2018). Japan Society of Traffic Engineers Journal of Traffic Engineering. Chiyoda, Tokyo, Japan: Japan Society of Traffic Engineers (JSTE). pp. 125-132

Chapter 9

Reinforcement Learning for Traffic Control Using Social Preferences

Orly Barzilai

Abstract

Traffic congestion arises from all directions, particularly during peak hours, and requires the implementation of a preference mechanism—designated lanes are set up as fast lanes for prioritizing public transportation and ride sharing. Defining a rigid criterion for using the fast lanes can be ineffective if the criterion for using these lanes is unrelated to traffic volume. In situations where fast lanes become overloaded, the rigid criteria do not ensure efficient travel. A social preference criterion, similar to those utilized in priority queues found in various service sectors such as government, travel, and cultural events, could be adapted for use in managing traffic flow and lane prioritization. The social preference criteria will be based on the driver's characteristics (e.g., a handicraft driver) or not its travel purpose (e.g., a doctor traveling for emergency surgery). To facilitate efficient travel for vehicles utilizing the fast lanes, the implementation of a reinforcement learning (RL) algorithm, specifically the Q-learning algorithm, is proposed. The results indicated that individuals exhibit social preference for various categories of vehicle passenger characteristics. The Q-learning algorithm regulated traffic flow in a junction simulation, distinguishing between fast lanes and regular lanes based on both social preference and traffic volume. This approach ensured efficient prioritization and allocation of resources.

Keywords: reinforcement learning (RL), traffic signal control (TSC), social preference, smart social junction (SSJ), fast lane (FL)

1. Introduction

In recent years, the growth of traffic congestion within urban areas has become a prominent and challenging issue, particularly in densely populated cities. Population increase, immense scale of vehicle trade, and lack of efficient public transportation systems are the foremost reasons for traffic saturation [1]. Traffic congestion is influenced by both spatial and temporal characteristics [2]. Different districts show different patterns of congestion. Traffic peaks during workdays typically occur in the morning and afternoon [2–5]. Peak traffic transpires in the morning between 8 am and 11 am, with another surge between 2 pm and 6 pm during the traditional workweek. The regularity of this pattern on workdays is a result of individuals collectively commuting to work at similar times throughout the day [3].

One of the primary challenges in populated cities arises at signalized junctions [4]. The predominant traffic control method globally is the fixed-cycle program, primarily due to the lack of real-time data availability. In this approach, the traffic light phases switch at predefined intervals. Nevertheless, the advent of wireless communication in recent years has brought about a significant change. Advancements in technology now allow for the acquisition of real-time road data. As a result, new traffic signal control (TSC) algorithms have been emerged utilizing data from a variety of devices including sensors, cameras, and monitors [6–12]. These devices facilitate the measurement, modeling, and interpretation of traffic features, including flow, occupancy, or travel times. This information proves valuable for the development of intelligent transportation systems (ITS) that manage traffic based on traffic density [13]. Among these algorithms, reinforcement learning (RL), an unsupervised machine learning technique, has been widely applied to develop solutions for optimizing TSC optimization [14, 15] and plays a central role in the field of traffic management [16–18].

Managing traffic signals based on traffic density can help alleviate traffic congestion when there are disparities in the number of vehicles arriving at junctions from various directions and heading in different ways. However, during peak hours when junctions are overwhelmed, it becomes essential to introduce a preference mechanism for distributing the traffic load over time [19]. Currently, a priority mechanism is implemented at junctions and on roads, offering preference to public transportation and high-occupancy vehicles. This is achieved through the establishment of dedicated lanes for these specific vehicle types, commonly referred to as dedicated bus lanes (DBL) and high-occupancy vehicles (HOV) lanes.

Barzilai et al. [20] suggested incorporating an innovative preference system based on the social characteristics of the vehicle driver (e.g., a handicapped person) or on its travel purpose (e.g., A doctor traveling to an emergency surgery at a hospital). The concept involves adjusting traffic signals according to both traffic load and social priority. The author argues that introducing a social preference parameter to the junction traffic management algorithm can reduce driver stress, by making traffic congestion at the intersection seem more just, consequently lowering accident rates. Another rationale for incorporating social preference is its integration with traffic load, creating a flexible mechanism for effectively managing traffic on dedicated lanes and preventing issues of both overloading and underloading [19].

In the following sections, we will elaborate on and summarize several papers that enhance the innovative concept of junction management, combining social priority with traffic load through the application of RL algorithms. The review of these papers will be complemented by an examination of the supporting literature.

2. Social preference as a criterion for traffic control

Moral reasoning is defined as "behavior that is subject to (or judged according to) generally accepted moral norms of behavior" [21]. Research on moral dilemmas' judgments has been fundamentally shaped by the distinction between Utilitarianism and Deontology. Gawronski et al. [22] suggests that the difference between these two concepts is based on sensitivity to moral consequences (Utilitarianism) and moral norms (Deontology). Lower levels of moral reasoning were found to be related to a higher number of accidents, to a higher driving speed, and to a higher degree of space-taking behavior [23].

Encountering a traffic jam, whether at a standstill in an intersection or within a lane, can be associated with the broader phenomenon of waiting in a line or in a queue. Queue involves waiting. Research into the impact of waiting time on consumers highlights both decreased satisfaction [24–26] and emotional effects such as stress, anxiety, boredom, and a sense of injustice. Moreover, perceived wait times often exceed reality, exacerbating the frustration of those affected [25–28].

Queues are identified as social structures that uphold specific social [29]:

- The queue visibly indicates the order of arrival.
- No skipping or cutting in line is allowed.
- Adherence to the "first come, first served" principle.
- Everyone must wait for their turn patiently.

Most people agree that deviations from queue norms, even in shorter lines, may be acceptable to accommodate concerns such as health, disability, or childcare, as long as the exception is requested and granted by fellow queue members. One's attitudes toward queues may be influenced by social injustice [30]. Queues are identified as social systems where social justice is an important factor for queue compliance. Social justice is defined by Rawls [31] as "fairness. The way in which social institutions distribute fundamental rights and duties and determine the division of advantages from social cooperation." On an individual level, social justice relates to "equitable and fair access to resources, and socially valued commodities" [32]. First-order justice, defined as a first-come, first-served (FCFS) process, has been found to be a necessary condition of social justice and positive evaluation [33]. This general principle is applied with modifications [34]. Second-order justice, defined as equal waiting time, has been found to be an additional factor that comes into play only when first-order justice is met. This principle is related to the volume of the service and cost, consumer's age, and physical conditions [33].

Priority queues are prevalent in-service operations, assigning priorities based on customer attributes [35]. Priority queues and their impact are investigated in airlines, theme parks, nightclubs, hotels, and other service contexts [36]. For example, in COVID-19 testing, priorities can be based on symptoms [37]. In government services, precedence can be granted to individuals who come from distant [38]. In numerous other scenarios, such customer characteristics are inaccessible, necessitating self-selection of priorities. This warrants the implementation of a pay-for-priority system, wherein customers who pay a higher fee are afforded greater precedence. Theme parks and ski resorts often allow customers to purchase premium tickets to join express lines. Everything Everywhere (EE), a leading telecommunications company in the United Kingdom, once offered "Priority Answer" that enabled customers to pay £0.50 to jump the queue for a service call [35].

Barzilai et al. [39] explored public sentiments related to the notion of prioritizing intersections using a rating scale from 1 (indicating strong opposition) to 5 (indicating strong agreement). Furthermore, insights into social preferences regarding various compositions of vehicle passengers were obtained through participants' ratings on a scale ranging from 1 (indicating no preference) to 5 (indicating a very strong preference). Social preference cases were categorized based on the distinction between moral principles (Deontology) and moral consequences (Utilitarianism).

Moral principles contained the following categories:

- Good virtues: people with special needs, elderly (over 75), pregnant women (8 months and above), and patients with active cancer.
- Benefit for all: public transportation (bus and cab), shared vehicles, and vehicles with four passengers and above.
- Encouraging community services: medical, repair technicians, security, education, and public services.
- Keeping traffic rules: normal speed, avoiding phone talk, and no insurance claim.
- Promotion of national issues: education, security, and health.

Moral consequences contained the following categories:

- Benefit for the individual: private bus and special cab.
- Promoting employment: driving to work (governmental and high-tech).
- Leisure time promotion: driving for leisure and volunteering.

The results revealed that while research participants advocate for equal priority for all vehicles (excluding emergency ones) at traffic junctions, they argue that priority adjustments should consider traffic volume. The participants preferred moral principles over moral consequences. Two categories received high preference. The first category is the benefit for all, which contains public and shared transportation.

In recent years, there has been a growing public awareness of the need to reduce road congestion through public and cooperative transportation. Barzilai et al. [39] suggested that people are affected by this public propaganda when selecting social preferences. The second category and most surprising finding was that the value of good virtues was associated with the highest preferences. People indicated that vehicles that contain people with special needs and difficulties should have higher priorities in a smart junction. The authors concluded that these results suggest that people can accept moral norms as a criterion for traffic load management.

3. Traffic control preference via dedicated lanes

Traffic control preference is implemented to prioritize public and ride-sharing transportation. Busses play a crucial role in public transportation. However, as busses share the road with private vehicles, they often encounter heavy traffic congestion and delays. To address this problem, bus priority solutions via bus lanes have evolved into an essential component of urban public transportation networks' growth and enhancement [40]. Russo et al. [41] describe two implementations for DBL systems. The first implementation involves the creation of extensive bus lanes that traverse the entire city. The second implementation consists of dedicated sections that are specifically allocated to address severe congestion within the bus network. These bus sections can be relatively short, sometimes spanning just a few hundred meters, with most of the bus routes operating on lanes not exclusively designated for them. Introducing or expanding such short bus sections constitutes a minor, localized modification to the city's transportation system, leading to minimal implementation costs. Montero-Lamas et al. [42] describe another bus priority solution known as an intermittent bus lane (IBL), where the bus lane's status is dynamic. The lane's status is changed to a mixed-traffic lane when there is no bus using it or when traffic conditions do not cause delays for the bus. When a typical traffic lane is transformed into a bus lane, it often leads to a decrease in the capacity of the remaining lanes, causing an increase in traffic congestion and a reduced travel speed on these capacity-reduced general lanes [43].

High-occupancy vehicle (HOV) lanes are specialized traffic lanes set for vehicles carrying multiple passengers. Depending on the number of occupants, a driver must either use the typically more congested general-purpose lane at a slower pace or have the option to access the faster-moving HOV lane. The fundamental goal of HOV lanes is to raise the average vehicle occupancy, thereby mitigating road congestion [44]. In most applications, these lanes require at least one or two passengers to accompany the driver. HOV lanes facilitate more efficient use of roadways, which benefits traffic flow while also providing time savings and enhanced reliability for high-occupancy travel modes [45]. Sometimes, HOV lanes are also made available to other vehicle types, including emergency and law enforcement vehicles, public busses, electric vehicles, or single-occupancy cars that choose to pay a toll [44].

To regulate the number of single-occupancy vehicles using the HOV lanes, an effective lane management strategy is the implementation of high-occupancy-toll lane pricing. HOV lane pricing can take one of three forms: (i) flat rates, which remain consistent over time; (ii) scheduled tolls, where fees change according to a predetermined schedule, such as the day of the week and time of day; or (iii) dynamic tolls, which are real-time and responsive to the current traffic conditions, ideally adjusted to the prevailing congestion levels [46]. In the case of scheduled tolls, elevated prices are implemented during periods of peak demand and the specific travel location [47]. An example could involve tolls calculated using a function that depends on the maximum traffic density downstream from the entry point [48]. For dynamic tolls, pricing should be responsive to demand. To initiate the implementation of HOV lane pricing, it is crucial that the demand is substantial enough to justify charging single-occupancy vehicles for their use [49]. Additionally, lane capacity should be considered during the planning process [50]. Numerous strategies have been developed to implement dynamic tolls, including the application of a multi-agent reinforcement learning algorithm [51] and a deep reinforcement learning algorithm [52]. A comprehensive understanding of dynamic toll pricing strategies and models is provided by Lombardi et al. [48].

4. Smart social junction with social preferences

Several studies employed a smart social junction (SSJ) algorithm where traffic light timings are adjusted based on both traffic load and societal preferences [19, 20, 39, 53]. To facilitate the implementation of this algorithm, the concept of conflict side (CS) was introduced. A CS encompasses all lanes in all directions of a traffic junction that can move simultaneously. A junction simulation was conducted with four CSs, as illustrated in **Figure 1**. Vehicles with randomly selected length

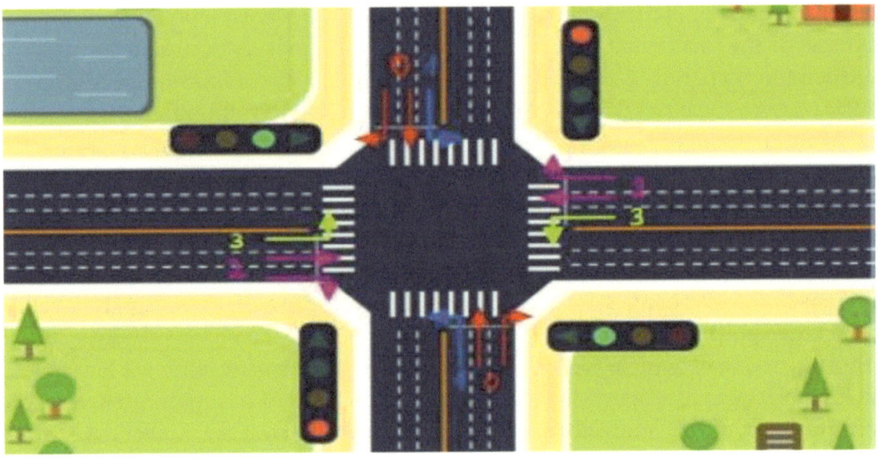

Figure 1.
The smart junction simulation's scheme with four conflict sides colored red, blue, pink, and green.

and velocity parameters were dispersed across lanes. Each vehicle was randomly assigned social preferences ranging from 1 to 10. The social preference value for each CS is the total of all individual vehicle preferences within that CS. Time slots are allocated to all CSs per cycle based on a constant scheduler, ensuring the same order is maintained in each cycle.

Figure 2 illustrates three different types of junctions examined in Barzilai et al. [39]: the standard junction, which allocates equal time slots to every CS; the SJ, which allocates timing slots to each CS based solely on its traffic load; and the SSJ, which allocates timing slots based on traffic load and social preference. The results indicate that shorter light durations were observed for the SMJ, although the timing of the SJ and SMJ were comparable for three out of four CSs.

Fine et al. [53] elaborated on the SSJ algorithm where, in each time slot, the CS with the highest preference value is selected. To prevent situations where certain CSs experience prolonged waits (starvation), additional points were allocated to a CS

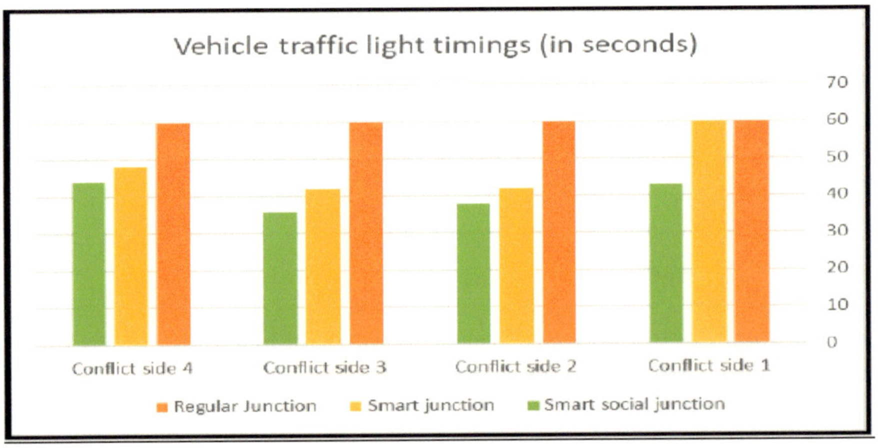

Figure 2.
Our model's timing in comparison with other traffic light timing methods.

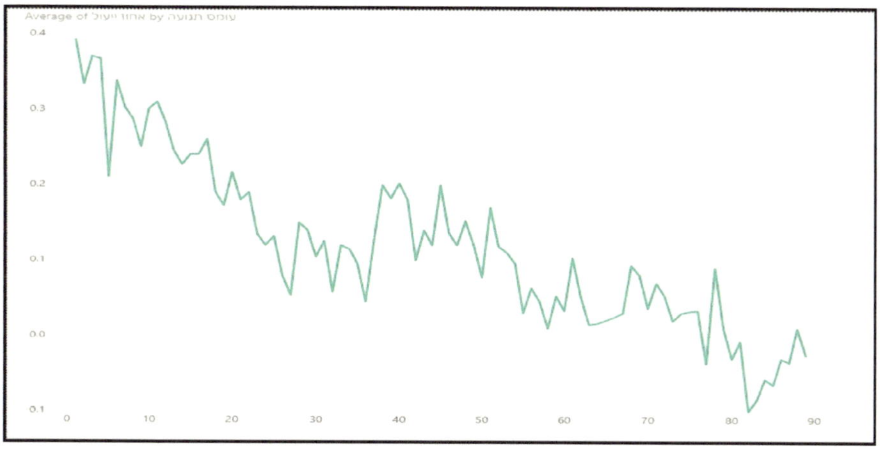

Figure 3.
Average efficiency of SSJ by traffic load.

that had not received a green light beyond a specified maximum waiting time. The researchers discovered that the SSJ outperformed the standard (regular) junction during low traffic volume. **Figure 3** depicts how the average efficiency of the SSJ evacuation changes in response to different traffic loads. As the traffic load increased, the efficiency declined. In addition, at a traffic load of 55%, the efficiency of both types of junctions (SSJ and regular junction) was comparable. When traffic load surpassed this threshold, regular junction proved more effective than their smart counterparts.

The researchers concluded that when all CSs experience the same level of congestion, there is no logical basis for prioritizing one side over another. Furthermore, the prioritization of individual vehicles is impacted by the presence of other vehicles in the same queue. For instance, if a high-priority vehicle is waiting alongside low-priority vehicles, the overall preference of the lane will be diminished, and the high-priority vehicle will not be able to assert its preference.

To address these challenges, Barzilai et al. [19] proposed implementing the concept of social preferences within fast lanes (FLs). Presently, FLs are designated for public transportation and are unaffected by traffic load. Consequently, traffic flow in these lanes often falls short of optimal levels, either due to overcrowding or insufficient volume. The proposed solution involves implementing adaptable FLs, where the types of vehicles permitted to use them vary based on prevailing traffic conditions. During periods of congestion and heightened demand, only vehicles with high social priority, such as public transportation, would be granted access to the FL. Conversely, when the FL is underused, vehicles with lower social priority could also utilize it. The optimization of traffic control in a FL based on social preference and traffic volume can be performed using the reinforcement learning (RL) algorithm.

5. Reinforcement learning algorithm for traffic control

Reinforcement learning (RL) is the problem faced by an agent that must learn behavior through trial-and-error interactions with a dynamic environment [54]. As a learning problem, it refers to learning to control a system to optimize some numerical values that represent long-term objectives [55]. The numerical values are the results of a reward function [56]. The reward function defines what is objectively good and bad for the agent. The value of the reward function is the agent's current mapping from the set of possible states (or state-action pairs) to its estimates of the net long-term reward to be expected after visiting a state (or a state-action pair), after which the agent continues to act according to its current policy [56]. Moreover, the learner is not told which actions should be taken, as in many forms of machine learning, but instead must discover which actions yield the highest reward by trying them out [56].

Q-learning algorithm is a model-free RL algorithm that does not rely on an explicit model of the environment. In model-free reinforcement learning, the agent learns to make decisions and improve its behavior through trial and error without building an internal representation or model of how the environment works [17]. In the Q-learning algorithm, the action is based on the maximal value of the received reward calculated per each action. Miletić et al. [17] introduced a schema for applying the Q-learning algorithm to traffic management at junctions. This schema involves employing a two-dimensional Q-table, where one dimension represents the states of the junction, and the other dimension corresponds to the actions taken at the junction.

The utilization of RL for traffic signal control (TSC) has gained popularity due to its capacity to learn and adapt while actively engaging with the environment. This inherent capability empowers the system to effectively respond to new patterns of traffic congestion as it encounters them [17]. In RL-TSC, each smart junction is typically controlled by a single agent [13]. Each agent is responsible for determining the light-switching sequence at its assigned junction. Many different objectives have been considered by authors when defining the reward function used by RL-TSC agents [13]. These may include average trip waiting time, trip delay, average trip time, average junction waiting time, junction throughput/flow rate, achieving green waves, accident avoidance, speed restriction, fuel conservation, and average number of trip stops.

6. Reinforcement learning algorithm for smart social junction

Barzilai et al. [19] utilized a simplified junction model consisting of two lanes: one for regular vehicles and another for priority vehicles, designated as FL. To effectively manage the distribution of green light time between these two lanes, ensuring priority for the FL without neglecting the regular lane, the RL algorithm, specifically the Q-learning algorithm, was employed. The Q-learning algorithm pseudo-code is presented in **Figure 4**. Their study applied random data, and the results demonstrated the success of applying the Q-learning algorithm. The reward function contained positive factors for vehicles that crossed the junction or advanced their position and a negative factor for vehicles that remained in their position. In addition, a weight value for the vehicles with high priority was also part of the equation. By implementing this algorithm, vehicles traveling in the FL were able to cross the junction more quickly than those in the regular lane, thus optimizing traffic flow and prioritizing vehicles based on social considerations.

In a recent study, Barzilai et al. [57] extended the simplified model of SSJ with FL managed by RL to a more realistic and practical solution. To achieve this, the junction simulation was constructed using actual data obtained from surveillance cameras situated at a complex, real-world junction. The surveillance camera data was extracted from "Netivei Israel" (the Israeli National Company for Transport Infrastructures) website for the "Ahisemech" junction (presented in **Figure 5**) which is located in the central region of Israel.

Pseudo code of the Q-Learning Algorithm:

1. Create a table Q of size $\#states \times \#actions$ and set all its entries to 0.
2. Create a vector of size $\#states$ and keep the reward of each state.
3. Runs N episodes (an iteration):
    a. For each episode:
        i. Given a current state s calculate and update the new values $Q(s, a_1)$ and $Q(s, a_2)$ in table Q, where a_1 and a_2 stand for the 2 possible actions (horizontal and vertical).
        ii. with ϵ probability, choose randomly the next action.
        iii. Otherwise, i.e. with $1 - \epsilon$ probability, choose the action a_i which maximize the Q-value, i.e. $argmax_{i=1,2}\{Q(s, a_1), Q(s, a_2)\}$
        iv. Moving to the next state s' by executing the selected action on the current state s.

Figure 4.
Pseudo-code of the Q-learning algorithm.

Figure 5.
Ahisemech junction photograph.

During the peak traffic hours between 8:00 am and 5:00 pm, a total of 40 minutes of video footage was captured, encompassing 2177 vehicles across 14 complete cycles of green light distribution in all directions. The researchers employed the You Only Look Once (YOLO) framework, specifically version 8, for conducting vehicle counts. YOLOv8 is a vision model utilized for object detection, classification, and segmentation tasks, particularly on real traffic videos serving diverse purposes [58–61].

Using the vehicle counts and the Ahisemech junction structure as reference points, a simulation was constructed. This simulation replicated the junction structure and CSs, as presented in **Figure 6**, as well as accurately represented the traffic load, as presented in **Table 1**. CS5 emerged as the preferred choice for prioritizing vehicles with high-priority status because it does not share lanes with other CSs.

Applying the Q-learning algorithm to manage the real data streamed to the smart junction simulation showed that the algorithm gave preference to both CS5, which was the least congested CS but was designated as the social priority CS, and CS2, which was

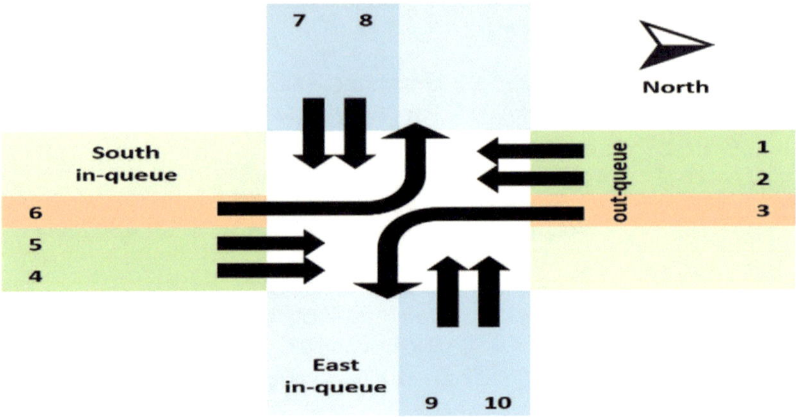

Figure 6.
Simulation of Ahisemech junction by lanes and CSs.

Lane	CS1	CS2	CS3	CS4	CS5	Total
L1 + L2		V	V			444
L3			V			6
L4 + L5	V	V				373
L6	V					62
L7					V	16
L8					V	81
L9				V		18
L10				V		103
Total	435	817	450	121	97	1103
Percentage	0.23	0.43	0.23	0.06	0.05	

Table 1.
Number of simulation vehicles by lane and by CS.

the most congested CS with no social priority. To assess the order of green light allocation, the average position for each CS was computed. This position reflects the timing of green light allocation. Green lights were assigned to various CSs, leading to a sequence of CS openings until the junction was completely evacuated. A lower value of the average position for a particular CS indicates that the green light was granted earlier in this sequence, resulting in a faster evacuation of that CS. The prioritization of CS5 and CS2 is evident from their low average position values. CS5 obtained a value of 13.4, constituting 64% of the total average position, whereas CS2 achieved a value of 6, accounting for 29% of the total average position. In addition, during the learning phase performed during the 35 episodes, a reduction of 30% was found for the preferred CS (CS5).

7. Conclusions

In response to the growing challenge of traffic congestion, there is a proposal for an innovative approach to implement a social priority mechanism within Traffic

Signal Control (TSC). This mechanism draws inspiration from various sectors, such as governmental, commercial, and healthcare domains, aiming to be integrated into the traffic control area.

A social preference based on driver characteristics or travel purpose is suggested, combined with the current traffic volume. This social preference is suggested to be implemented through dedicated lanes designated as fast lanes (FLs). To effectively manage the traffic flow between regular lanes and fast lanes, the reinforcement learning (RL) algorithm, specifically the Q-learning algorithm, is suggested, enabling a flexible usage of the fast lane depending on the current traffic load. Simulation of a junction, imitating the structure and traffic volume of an actual junction, has demonstrated the effectiveness of using RL to optimize traffic balance, considering both load and social preferences.

The implementation of an algorithm in a real-life traffic junction or lanes could be implemented through a smartphone or smart car software that connects to road sensors. Validating social preferences more proficiently can be done in the form of biometric authentication. The validation itself could be implemented in smart calendar applications, in which a certain institute, such as a hospital or government office, can provide schedule validation for meetings and appointments.

Future research directions can focus on refining the social priority categories, enhancing the algorithm to encompass more complex environments connecting several junctions, and inferring social priority designated to a vehicle in real time.

Author details

Orly Barzilai
The Academic College of Tel Aviv-Yaffo, Tel Aviv-Jaffa, Israel

*Address all correspondence to: orlyba@mta.ac.il

References

[1] Rizwan P, Suresh K, Babu MR. Real-time smart traffic management system for smart cities by using Internet of Things and big data. In: 2016 International Conference on Emerging Technological Trends (ICETT). Kollam, India: IEEE; 2016. pp. 1-7

[2] Li X, Gui J, Liu J. Data-driven traffic congestion patterns analysis: A case of Beijing. Journal of Ambient Intelligence and Humanized Computing. 2023;**14**(7):9035-9048

[3] Almatar KM. Traffic congestion patterns in the urban road network: (Dammam metropolitan area). Ain Shams Engineering Journal. 2023;**14**(3):101886

[4] Kolat M, Kővári B, Bécsi T, Aradi S. Multi-agent reinforcement learning for traffic signal control: A cooperative approach. Sustainability. 2023;**15**(4):3479

[5] Zang J, Jiao P, Liu S, Zhang X, Song G, Yu L. Identifying traffic congestion patterns of urban road network based on traffic performance index. Sustainability. 2023;**15**(2):948

[6] Al-Abaid SAF. A smart traffic control system using image processing: A review. Journal of Southwest Jiaotong University. 2020;**55**(1):1-6

[7] Ata A, Khan MA, Abbas S, Ahmad G, Fatima A. Modelling smart road traffic congestion control system using machine learning techniques. Neural Network World. 2019;**29**(2):99-110

[8] Chong HF, Ng DWK. Development of IoT device for traffic management system. In: 2016 IEEE Student Conference on Research and Development (SCOReD). Kuala Lumpur, Malaysia: IEEE; 2016. pp. 1-6

[9] Firdous A, Niranjan V. Smart density based traffic light system. In: 2020 8th International Conference on Reliability, Infocom Technologies and Optimization (Trends and Future Directions) (ICRITO). Noida, India: IEEE; 2020. pp. 497-500

[10] Hartanti D, Aziza RN, Siswipraptini PC. Optimization of smart traffic lights to prevent traffic congestion using fuzzy logic. TELKOMNIKA (Telecommunication Computing Electronics and Control). 2019;**17**(1):320-327

[11] Lana I, Del Ser J, Velez M, Vlahogianni EI. Road traffic forecasting: Recent advances and new challenges. IEEE Intelligent Transportation Systems Magazine. 2018;**10**(2):93-109

[12] Patil PG, Sharma S, Tamilisetti C, Prathap S. Real time smart traffic control system. International Journal of Research in Engineering, Science and Management. 2020;**3**(2):1-4

[13] Mannion P, Duggan J, Howley E. Parallel reinforcement learning for traffic signal control. Procedia Computer Science. 2015;**52**:956-961

[14] Arel I, Liu C, Urbanik T, Kohls AG. Reinforcement learning-based multi-agent system for network traffic signal control. IET Intelligent Transport Systems. 2010;**4**(2):128-135

[15] Balaji PG, German X, Srinivasan D. Urban traffic signal control using reinforcement learning agents. IET Intelligent Transport Systems. 2010;**4**(3):177-188

[16] Haydari A, Yılmaz Y. Deep reinforcement learning for intelligent transportation systems: A survey. IEEE Transactions on Intelligent Transportation Systems. 2020;**23**(1):11-32

[17] Miletić M, Ivanjko E, Gregurić M, Kušić K. A review of reinforcement learning applications in adaptive traffic signal control. IET Intelligent Transport Systems. 2020;**16**(10):1269-1285

[18] Noaeen M, Naik A, Goodman L, Crebo J, Abrar T, Abad ZSH, et al. Reinforcement learning in urban network traffic signal control: A systematic literature review. Expert Systems with Applications. 2022;**199**:116830

[19] Barzilai O, Rika H, Voloch N, Hajaj MM, Steiner OL, Ahituv N. Using machine learning techniques to incorporate social priorities in traffic monitoring in a junction with a fast lane. Transport and Telecommunication Journal. 2023;**24**(1):1-12

[20] Barzilai O, Voloch N, Hasgall A, Lavi Steiner O, Ahituv N. Traffic control in a smart intersection by an algorithm with social priorities. Contemporary Engineering Sciences. 2018;**11**(31):1499-1511

[21] Reynolds SJ, Ceranic TL. The effects of moral judgment and moral identity on moral behavior: An empirical examination of the moral individual. Journal of Applied Psychology. 2007;**92**(6):1610-1624. DOI: 10.1037/0021-9010.92.6.1610

[22] Gawronski B, Armstrong J, Conway P, Friesdorf R, Hütter M. Consequences, norms, and generalized inaction in moral dilemmas: The CNI model of moral decision-making. Journal of Personality and Social Psychology. 2017;**113**(3):343-376

[23] Veldscholten N. Moral Reasoning in Traffic: About the Possible Relations Between Moral Reasoning and Traffic Safety [Master's thesis]. University of Twente; 2015

[24] Chang HL, Yang CH. Do airline self-service check-in kiosks meet the needs of passengers? Tourism Management. 2008;**29**(5):980-993

[25] Carmon Z, Shanthikumar J, Carmon T. A psychological perspective on service segmentation models: The significance of accounting for consumers' perceptions of waiting and service. Management Science. 1995;**41**(11):1806-1815

[26] Hirsh I, Bilger R, Deatherage B. The effect of auditory and visual background on apparent duration. The American Journal of Psychology. 1956;**69**(4):561-574

[27] Maister D. The psychology of waiting lines. In: Czepiel J, Solomon M, Surprenant C, editors. The Service Encounter: Managing Employee/Customer Interaction in Service Businesses. Lexington, MA: Lexington Books; 1985

[28] Witowska J, Schmidt S, Wittmann M. What happens while waiting? How self-regulation affects boredom and subjective time during a real waiting situation. Acta Psychologica. 2020;**205**:103061

[29] Fagundes D. The social norms of waiting in line. Law & Social Inquiry. 2017;**42**(4):1179-1207

[30] Larson RC. Perspectives on queues: Social justice and the psychology of queueing. Operations Research. 1987;**35**(6):895-905. DOI: 10.1287/opre.35.6.895

[31] Rawls J. A Theory of Justice. Cambridge: Belknap Press of Harvard University Press; 1971

[32] Zajda J, Majhanovich S, Rust V. Education and Social Justice. Heidelberg, Germany: Springer Verlag; 2006

[33] Chatterjee S. Order of justice in queues of emerging markets. Journal of Consumer Marketing. 2020;**37**(6):605-616

[34] Brady FN. Lining up for Star-Wars tickets: Some ruminations on ethics and economics based on an internet study of behavior in queues. Journal of Business Ethics. 2002;**38**:157-165

[35] Cui S, Wang Z, Yang L. A brief review of research on priority queues with self-interested customers. In: Innovative Priority Mechanisms in Service Operations: Theory and Applications. Heidelberg, Germany: Springer; 2023. pp. 1-8

[36] Alexander M, MacLaren A, O'Gorman K, White C. Priority queues: Where social justice and equity collide. Tourism Management. 2012;**33**(4):875-884

[37] Yang L, Cui S, Wang Z. Design of covid-19 testing queues. Production and Operations Management. 2022;**31**(5):2204-2221

[38] Wang Z, Cui S, Fang L. Distance-based service priority: An innovative mechanism to increase system throughput and social welfare. Manufacturing and Service Operations Management. 2022;**25**(1):353-369

[39] Barzilai O, Voloch N, Hasgall A, Steiner OL. Real life applicative timing algorithm for a smart junction with social priorities and multiple parameters. In: 2018 IEEE International Conference on the Science of Electrical Engineering in Israel (ICSEE). Eilat, Israel: IEEE; 2018. pp. 1-5

[40] Xue Y, Cheng L, Zhong M, Sun X. Evaluation of bus lane layouts based on a bi-level programming model—Using part of the Qingshan Lake District of Nanchang City, China, as an example. Sustainability. 2023;**15**(11):8866

[41] Russo A, Adler MW, van Ommeren JN. Dedicated bus lanes, bus speed and traffic congestion in Rome. Transportation Research Part A: Policy and Practice. 2022;**160**:298-310

[42] Montero-Lamas Y, Novales M, Orro A, Currie G. A new big data approach to understanding general traffic impacts on bus passenger delays. Journal of Advanced Transportation. 2023:1-15

[43] Kampouri A, Politis I, Georgiadis G. A system-optimum approach for bus lanes dynamically activated by road traffic. Research in Transportation Economics. 2022;**92**:101075

[44] Boysen N, Briskorn D, Schwerdfeger S, Stephan K. Optimizing carpool formation along high-occupancy vehicle lanes. European Journal of Operational Research. 2021;**293**(3):1097-1112

[45] Gitelman V, Doveh E. Examining the safety impacts of high-occupancy vehicle lanes: International experience and an evaluation of first operation in Israel. Sustainability. 2023;**15**(18):13976

[46] De Palma A, Lindsey R. Traffic congestion pricing methodologies and technologies. Transportation Research Part C: Emerging Technologies. 2011;**19**(6):1377-1399

[47] DeCorla-Souza P. Making the pricing of currently free highway lanes

acceptable to the public. Transportation Quarterly. 2000;**54**(3):17-20

[48] Lombardi C, Picado-Santos L, Annaswamy AM. Model-based dynamic toll pricing: An overview. Applied Sciences. 2021;**11**(11):4778

[49] Martínez I, Jin WL. Dynamic Distance-Based Pricing Scheme for High-Occupancy-Toll Lanes Along a Freeway Corridor. arXiv preprint arXiv:2309.01990. 2023

[50] Pulyassary H, Yang R, Zhang Z, Wu M. Capacity Allocation and Pricing of High Occupancy Toll Lane Systems with Heterogeneous Travelers. arXiv preprint arXiv:2304.09234. 2023

[51] Pandey V, Boyles SD. Multiagent reinforcement learning algorithm for distributed dynamic pricing of managed lanes. In: Proceedings of the IEEE Conference on Intelligent Transportation Systems, Proceedings, ITSC, Maui, HI, USA. IEEE; 4-7 Nov 2018. pp. 2346-2351

[52] Pandey V, Wang E, Boyles SD. Deep reinforcement learning algorithm for dynamic pricing of express lanes with multiple access locations. Transportation Research Part C: Emerging Technologies. 2020;**119**:102715

[53] Fine Z, Brayer E, Proshtisky I, Barzilai O, Voloch N, Steiner OL. Handling traffic loads in a smart junction by social priorities. In: 2019 IEEE International Conference on Microwaves, Antennas, Communications and Electronic Systems; November 2019: (COMCAS). Israel Tel Aviv: IEEE; 2019. pp. 1-5

[54] Kaelbling LP, Littman ML, Moore AW. Reinforcement learning: A survey. Journal of Artificial Intelligence Research. 1996;**4**:237-285

[55] Szepesvári C. Algorithms for reinforcement learning. In: Synthesis Lectures on Artificial Intelligence and Machine Learning 4.1. Heidelberg, Germany: Springer; 2010. pp. 1-103

[56] Sutton RS, Barto AG. Reinforcement learning. Journal of Cognitive Neuroscience. 1999;**11**(1):126-134

[57] Barzilai O, Rika H, Hassin Y. Smart social junction traffic control using reinforcement learning on real data. IET Intelligent Transport Systems. 2024

[58] Chaudhary HK, Saraswat K, Yadav H, Puri H, Mishra AR, Chauhan SS. A Real Time Dynamic Approach for Management of Vehicle Generated Traffic. Transdisciplinary Research and Education Center for Green Technologies, Kyushu University; 2023. pp. 289-299

[59] Gomaa A, Minematsu T, Abdelwahab MM, Abo-Zahhad M, Taniguchi RI. Faster CNN-based vehicle detection and counting strategy for fixed camera scenes. Multimedia Tools and Applications. 2022;**81**(18):25443-25471

[60] Zhang Y, Guo Z, Wu J, Tian Y, Tang H, Guo X. Real-time vehicle detection based on improved yolo v5. Sustainability. 2022;**14**(19):12274

[61] Rodríguez-Rangel H, Morales-Rosales LA, Imperial-Rojo R, Roman-Garay MA, Peralta-Peñuñuri GE, Lobato-Báez M. Analysis of statistical and artificial intelligence algorithms for real-time speed estimation based on vehicle detection with YOLO. Applied Sciences. 2022;**12**(6):2907

Chapter 10

Enhancing Road Safety in India: A Predictive Analysis Using Machine Learning Algorithm for Accident Severity Modeling

Humera Khanum, Rushikesh Kulkarni, Anshul Garg and Mir Iqbal Faheem

Abstract

This chapter presents a comprehensive study aimed at enhancing road safety in India through the development and application of a machine-learning predictive model for traffic accident severity on Indian highways. With road accidents being a leading cause of death and injury, claiming approximately 1.35 million lives globally each year, India faces a particularly acute challenge, reporting nearly 449,002 road accidents in 2019 alone. This work leverages the adaptability and superior predictive accuracy of machine-learning algorithms to model accident severity, thereby providing a basis for understanding contributing factors and formulating effective preventive strategies. Employing a meticulous multistep methodology, this study involves the collection and preparation of data from authorized organizations for data availability, feature selection, model training, parameter tuning, and model evaluation based on statistical accuracy matrixes. The chapter concludes by highlighting the significant potential of integrating machine-learning techniques with enhanced data recording systems to improve road safety modeling, decision-making, and accident prevention, ultimately contributing to the reduction of road traffic accidents and their associated human and economic costs.

Keywords: road safety in India, machine-learning algorithms, accident severity modeling, traffic accident analysis, predictive analytics, accident prevention strategies

1. Introduction

1.1 Background

Across the globe, the issue of road safety stands out as one of the most significant areas of public health and development. Unfortunately, India bears an unequal share of this burden [1]. It is worth noting that according to the Global Status Report on Road Safety 2018, a significant number of road accidents are a threat to life as well as general health, thereby calling for comprehensive and more effective interventions [2, 3]. Particularly, the Indian subcontinent is in a peculiarly fragile state as it relates to traffic accidents, that claim lives besides maiming many others involved [4]. It has arisen because of a myriad of factors like driver behavior among them a lack of tolerance for other road users.

This chapter explores how machine-learning algorithms can enhance road safety in India by building a robust model for predicting the severity of an accident. The primary purpose of this study is to identify what causes accidents to be severe using a large dataset and analyze relevant traffic and road characteristics as outlined in the chapter "The effect of traffic and road characteristics on road safety: A review and future research direction" [4]. Information from this analysis could be instrumental in guiding targeted interventions on road infrastructure improvement before setting up a secure transport system.

It is worrying that the question of road safety in India is facing a crisis situation now. India is currently grappling with a severe road safety crisis which largely contributes to worldwide road traffic fatalities. A high incidence of road accidents is reported due to its high and increasing population in addition to rising car numbers alongside poor roads that do not support them Hence, despite owning only 1% of the world vehicles population, India contributed 11% of global death from road accidents according to Global Status Report on Road Safety 2018 (World Health Organization) cited earlier [1]. Therefore, there is an urgent need for proper road safety mechanisms in such a nation.

Road accidents result from an interaction between different factors, which include human error such as speeding, alcohol drinking while driving, use of a phone during driving, or driver fatigue significantly leads to the occurrence of road crashes. Poor infrastructure like bad road designs, poor lighting systems nonexistence of some precautions like pavements or guardrails as well as inadequacy in maintenance also predisposes the happening of accidents along roads. Vehicle condition: malfunctions may cause accidents when they occur in a car; brake failure, and tire burst among others may cause accidents. Weather conditions like rain, fog, or snow reduce visibility making the roads more slippery hence increasing the chances of having accidents on such roads. Traffic conditions: this involves too many vehicles on the road, traffic jams as well as aggressive driving behaviors among motorists which may lead to many road accidents happening every day.

Traditional road safety measures: in the bid to see that there is a reduction in the occurrence of crashes, several measures have been put in place over the years. Some of these conventional road safety measures have borne fruits when it comes to minimizing accidents, while still there are many other methods that can be employed. However, it is not always easy to follow this path because they face some difficulties that are as follows:

Reactivate approach: conventional interventions are rather responsive, being concerned with dealing with incidents once they occur not preventing them.

Data limitations: there is little data for focused interventions due to a lack of adequate information regarding the accidents happening along the road.

Enforcement challenges: the enforcement of traffic rules may be difficult in many countries due to the high population and scarce resources particularly those in the developing world.

Predictive analytics together with machine-learning were born as an alternative approach due to the limitations of traditional road safety interventions [5, 6]. These technologies have the following capabilities:

Historical records on accidents could be used to detect patterns using machine-learning algorithms. Once patterns are detected, they can be used to formulate predictive models.

Machine learning programs can examine big sets of data to find factors that increase the risk of accidents by studying past accidents. To prevent accidents on roads, decision-makers need to be aware of the likelihood of accidents occurring beforehand. Machine Learning models can help improve how money is spent on enforcing laws, running awareness campaigns, and other efforts to reduce traffic deaths. They are also effective in deciding how to allocate resources for law enforcement, improving roads, and educating people about road safety using data as a source [7].

The main goal of this review was to construct a predictive model on different accident incidents' severities within India through the use of machine learning methods by Building Forecasting Models for Accident Severity.

To overcome the challenge of improving safety levels on our roads, this research adopts a machine-learning approach aimed at building a predictive model that can be used to predict accident severity in India.

This chapter attempts to address the pressing issue of road safety within India by exploring the application potential of machine learning in predicting the extent of injury caused during an accident. Consequently, it will draw from the original work—"The effect of traffic and road characteristics on road safety." This methodology would largely entail applying large datasets to pinpoint major drivers behind accidents' seriousness hence leading toward designing preventive interventions based on data analysis. The focus of this chapter is thus to provide information on specific interventions and road safety measures aimed at enhancing the transportation system's safety within India.

2. Literature review

This research derives from the existing body of knowledge about the road safety situation and utilizes machine-learning techniques, which give valuable insights.

Globally, road traffic accidents are an urgent public health problem that results in many deaths, injuries, and significant economic losses. India is currently grappling with a major road safety issue owing to rapid population growth coupled with a rise in the number of vehicles on its roads [2, 8].

Traditional road safety measures, though not completely worthless, have tended toward a reactive and somewhat restricted approach. In several cases, this has resulted in the development of innovative solutions based on data, with machine learning showing potential for enabling proactive interventions.

However, its ramifications extend globally, mostly impacting less developed nations that account for over 85% of victims worldwide but only receive half of the licenses dispensed annually.

Road accidents result from a complex interplay of several factors. An analysis of traffic characteristics, road infrastructure, and their interaction is provided which underscores how such factors influence the frequency and severity of accidents [4]. The research argues that efficient safety strategies must take into consideration factors like traffic volume, speed, road geometry, and infrastructure design.

Human error, which encompasses behaviors like speeding, drunk driving distracted driving or even 'driving while tired', is still a major factor in road traffic crashes. There is an intricate connection between driving behavior and road safety.

Thus, it adopts a theoretical framework that incorporates personal tendencies within society's expectations against which cultural norms govern driving conduct in different contexts. Thus, focusing interventions would be important in changing such dangerous driving behaviors to facilitate road safety policy formulation [9].

Improving road safety not only requires focusing on infrastructure and enforcement but also creating a safety culture among users of the road. In this regard, the need to incorporate road safety education in school curricula is underscored, as it equips young people with adequate information on how well they can use roads safely when using them. In addition, further highlights that point concerning targeted driver education programs which aim especially at professionals thus leading to better safety outcomes on roads.

Successful road safety strategies need sound institutional frameworks and policy interventions based on solid evidence. The study examined how institutions and policy instruments affected road safety promotion in Uganda, focusing on public health outcomes. According to the research, the use of collaborative approaches that involve stakeholders as well as making evidence-based decisions will lead to sustainable road safety improvement.

Advancements in technology particularly data science and machine learning present opportunities in improving road safety. According to data analysis plays a crucial role in pinpointing safety concerns within the road safety ecosystem. The study underlined the importance of collecting and analyzing large-scale information sets for purposeful action planning and stakeholders. Overall, machine learning offers a powerful means for anticipating the seriousness of highway accidents, which can result in both proactive safety measures and the identification of high-risk situations.

"Modelling Road Accident Severity with Comparisons of Logistic Regression, Decision Tree, and Random Forest" assesses the effectiveness of different machine learning techniques that include logistic regression, classification, and regression tree, as well as random forest in predicting road accident severity. In the study, it is shown that there is no other method that can produce more accurate results than the use of random forests in estimating the level of severity of an accident [10, 11]. The article argues on the ability of this technique in forecasting for accident severity perspective.

"Identification of potential traffic accident hot spots based on accident data and GIS" shows how important spatial attributes should be integrated with crash data at both macro and micro levels when developing prediction models.

3. Road safety in India: an overview

The planet continues to be anxious regarding road safety as a whole and India is among the countries with higher challenges concerning this aspect. Therefore, the high number of road traffic crashes and associated injuries serve as justification for the development of novel ways that aim at preventing these incidents from happening [12]. This chapter focuses on enhancing road safety in India through predictive analysis using machine learning algorithms for accident severity modeling.

India is one of the countries having the largest road networks globally and this leads to numerous accidents due to among other reasons, driver behavior [13], condition of roads, and vehicle-related causes. The Global Status Report by the World Health Organization on Road Safety points out that road traffic injuries are the leading cause of death for people aged 5–29 years globally. Specifically, India's Motor

Vehicle Accident Report by the Government reveals a worrying trend in the severity of road crashes and loss of lives.

However, machine intelligence is here to take care of such challenges to manage them effectively. In this regard, machine intelligence is able to sift through massive sums of data in order to establish patterns that can be helpful in predicting where an accident is more likely to happen, what causes it, and in the end the means of containing it. Some studies have endorsed the application of machine learning in traffic accident analysis and hotspot prediction using various techniques such as decision trees, random forests, logistic regression, and so on [14].

This chapter seeks to examine how machine learning algorithms can be used in predicting the seriousness of road accidents especially in India. Predictive analytics based on historical data is crucial for identifying factors associated with higher relative severity thereby allowing for identification of potential accident hotspots. The primary intent is not just about reducing these incidences but also about ensuring that they are less severe resulting in saving lives as well as minimizing economic losses from road traffic-related incidents.

It is important to acknowledge that this involves many different things and not only accident severity modeling using machine learning. Consequently, the possibility of predictive analytics within transforming strategies for road safety depicts a shift from being reactive to proactive. This chapter thus provides theoretical background to ML in analyzing accidents; reviews some applied literature on this subject; and introduces methods used by this science discipline alongside describing some real-life examples when such methods were applied within the Indian environment.

In short advancing road safety through predictive analytics using ML based on severity estimation appears as an avenue toward mitigating the heavy burden of these incidents on roads in India. By using data and advanced analytics, actors involved in these processes will develop more effective methods to prevent road accidents which make Indian roads and other areas safer.

4. Machine learning algorithms: a primer

Machine learning (ML) algorithms are driving the data science revolution by transforming various industries and academic inquiry in themselves since they can learn from data, identify patterns, and make decisions with minimal human intervention. Ultimately this makes them very useful for dealing with tough problems that cannot be solved through traditional analytic methods. Thus, the classifications of ML algorithms range from their foundations to their types to how they can be exploited in modeling accident severity for road safety.

In terms of how they learn, ML algorithms fall under three main categories.

Supervised learning: in supervised learning, these algorithms are given labeled datasets and they try to determine what each feature of the input means with regard to the output. Using this information, they can henceforth predict unseen data or make decisions based on new information acquired through training such as regression models including linear regression models; decision trees; and support vector machines among others.

Unsupervised learning: these are algorithms that do not use labeled observations but look for patterns within the data itself such as clustering techniques or dimensionality reduction methods. An example includes k-means clustering which is an unsupervised algorithm while principal component analysis (PCA) falls under this category.

Reinforcement learning: this type of machine learning is based on the idea that when a living organism gets feedback for its actions, the organism learns how to act in an environment.

Foundational concepts: there are some foundational concepts behind how ML algorithms work.

Feature selection: feature selection is about choosing the most relevant input features to use in predicting.

Model training: this is the process of teaching or training ML algorithms how to make predictions or decisions based on data whose outcomes are already known.

Parameter tuning: this involves adjusting the settings within the ML model so that more accurate results can be achieved.

Model evaluation: evaluating the performance of a machine learning model according to specific metrics such as accuracy, precision-recall curves (PRCs), F1 score for classification, or Mean Square Error (MSE) in regressions.

Applications in road safety: in the area of the road safety domain, machine-learning systems have been mainly employed to develop models that are capable of predicting accident severity levels. For example, based on historical accident data over time, these models can provide probabilities of future occurrences and their related injuries or deaths, hence enabling targeted measures. For instance, supervised learning algorithms could predict accident hotspots or severity levels through road status factors, including weather conditions and traffic volume, among many others. At the same time, unsupervised learning of hidden patterns in accident data, which would not be evident at first glance, can provide new insights into strategies for prevention.

To summarize, ML algorithms are highly effective at solving complex challenges across different sectors, especially in making roads safer. Considering this fact, they lead to a reduction in accident severity through learning from data while giving exact recommendations on how to enhance road safety in general or save lives in worst-case scenarios. Nonetheless, one might expect certain progressions regarding AI use within roads due to technological advancements, such as addressing traffic congestion through these machines thereby requiring necessary knowledge pertaining to ML algorithms together with their applications.

5. Methodology

The suggested approach for this investigation encompasses the subsequent stages of actualizing an RF technique for estimating accident intensity by making use of machine learning concepts.

1. Data preparation: the initial stage to avail an RF model for prediction on the severity of injuries is to collect all the data and get it prepared. The data on road accidents for specific road stretches can be sourced from other sources such as the Ministry of Road Transport and Highways (MoRTH) and the National Highways Authority of India (NHAI).

2. Data wrangling and mining techniques shall be used to clean and preprocess the data.

3. Feature selection: after you have completed your data preparation process, selecting appropriate features becomes essential. According to its definition,

feature selection refers to identifying and selecting the most useful predictors or independent variables. There are many ways to choose features, including statistical tests, correlations, and principal component analysis (PCA).

4. Model training: to this end, the preprocessed data can be used to train an RF model. Such a model can be created using machine learning. This algorithm uses bootstrap aggregating as well as random feature selection to build several decision trees that are then combined so as to attain better performance.

RF algorithm formulation: the RF algorithm can be represented as:

$$RF(X) = \frac{1}{B}\sum_{b=1}^{B} T_b(X) \tag{1}$$

where X is the input features, B is the number of trees, and $T_b(X)$ is the prediction of the b-th individual decision tree.

5. Parameter tuning: for enhancing the performance of a random forest model, it is critical to tweak its settings. The tuning performance of a random forest mainly depends on three key parameters: the overall number of trees (n_estimators), the number of features for the node splitting (max_feature), and the maximum tree depth (max_depth).

In creating an RF model used to predict traffic accident gravity, we usually use Gini impurity as one of the measures to evaluate the importance of different explanatory variables. At each node split, Gini impurity, a measure employed in decision trees that are base learners within the RF framework, remains critical for selecting features optimally. It is a quantitative measure that guides us in noting how effectively this variable distinguishes our target classes.

Mechanism of Gini impurity: in the context of binary classification, the Gini impurity for a node is calculated as:

$$I\_G(P) = \sum_{k=1}^{n} p^2 k \tag{2}$$

where P is the proportion of samples classified to class k at that node, and the summation operates over all classes. A lower Gini impurity score suggests a higher purity of the node, indicating an enhanced classification.

Gini importance in RF: Gini impurity serves two purposes in the developed RF model: node splitting and feature importance. Node splitting helps in identifying the most important variable at each node by checking how much purity can be reduced due to split on all potential splits, while feature importance computes Gini importance which is the average reduction of impurity caused by each feature after training all trees. Gini importance provides insights on which features have more weight in relation to making predictions in this task.

Model evaluation: it is important to evaluate the model's performance after training the RF model and optimizing its parameters. Different evaluation metrics like accuracy, precision, recall, F1 score, and Area Under The Curve-Receiver Operating Characteristics (AUC-ROC) curve may thus be employed.

Model implementation: after training and evaluating the model, it is ready to be used in predicting the seriousness of road accidents. This implies that a quantifiable process can be established with the aim of developing an algorithm in Python that will aid in predicting how bad road traffic will result in the future on highways in India.

6. Case study review: accident severity prediction

The study areas selected were the two stretches of Indian National Highways, (1) Pune-Sholapur Section of NH-9 in km 144/400 to km 249/000 in the State of Maharashtra, and (2) Six-Laning of Barwa-Adda-Panagarh Section of NH-2 from km 398.240 to km 521.120 including Panagarh Bypass in the States of Jharkhand and West Bengal [15] (**Figure 1**).

The study areas were selected based on specific criteria. Firstly, the researchers had prior experience working on one of the stretches, the Pune-Sholapur Section of NH-9, from km 144/400 to km 249/000 in the State of Maharashtra. This experience could have provided insights and knowledge that could be useful in conducting the study. Additionally, data were also provided by the same concessionaire as of the previous stretch on request for another stretch, which is the Six-Laning of Barwa-Adda-Panagarh Section of NH-2 from km 398.240 to km 521.120, including the Panagarh Bypass in the States of West Bengal. This data could have been relevant to the research objectives and could have assisted in achieving the desired outcomes.

The primary aim of the research was to create a predictive model for the severity of traffic accidents on Indian highways utilizing Random Forest models, chosen for their precision and comprehensibility. The study's results were employed to establish a predictive model for accident severity, which can contribute to the

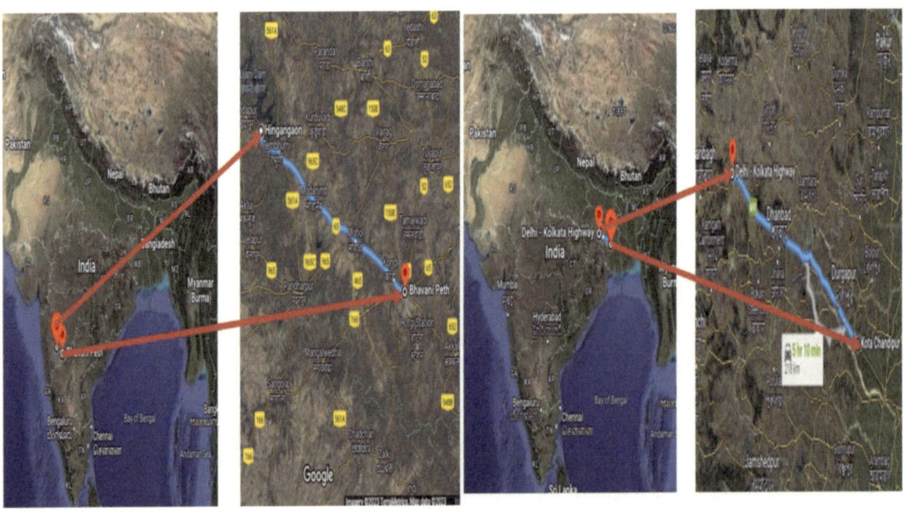

Figure 1.
Pune-Sholapur Section of NH-9 in the state of Maharashtra and Barwa-Adda-Panagarh Section of NH-2 in the states of Jharkhand and West Bengal.

formulation of road safety strategies and measures. This model enables the identification of high-risk zones and the allocation of resources for accident prevention and mitigation.

6.1 Data collection and preparation

Source data: the analysis focused on the Pune-Solapur Section of NH-9, covering accident records from 2013 to 2018, and the Six-Laning of Barwa-Adda-Panagarh Section of NH-2, encompassing accident data from 2015 to 2019. Data on road accidents was collected from the Concessionaires of the National Highways Authority of India for these projects. Subsequently, exploratory data analysis was conducted on the raw data.

Data preparation: the secondary source data were utilized for exploration. The dataset comprises 3257 observations, with 1855 observations of the Bengal (BAEL) Section and 1402 observations related to Pune-Solapur. It includes 32 variables, among which is the target variable "accident severity." **Table 1** displays the attributes and their respective mappings.

6.2 Data modeling

The RF classification algorithm has been employed in this study to forecast the severity of road traffic accidents in India. This section details the procedure for implementing the model and performance evaluation and discusses the results obtained.

The target variable for the RF model is selected as the accident Severity' which has classes as Fatal, Grevious Injury, Minor Injury and No Injury and indexed as [1—Fatal, 2—Grevious Injury, 3—Minor Injury, 4—No Injury]. The dataset is partitioned into training and testing sets with a ratio of 80 and 20%, respectively. The hyperparameters'n_estimators' and'max_depth' are specified, and a grid search is conducted with cross-validation (cv = 5) to identify the optimal hyperparameters. The best parameters and scores are obtained. The best estimator is fit on the training data. Predictions are made on the test data and the accuracy of the model is obtained. The algorithm and program for Accident Severity Modeling using RF is written in the Python programming language, and the code is made available to the public for further development. The source code can be accessed *via* the software availability statement. Accuracy analysis on test data: Three metrics were employed to evaluate the effectiveness of the algorithms: accuracy, precision, and recall.

6.3 Result and discussion

Model performance: The model used three hyperparameters: 'max_depth':10, 'max_features':'sqrt', and 'n_estimators': 100. The confusion matrix showed correct and incorrect classifications per class. With support, the classification report displayed precision, recall, and f1-score per class. The model had high precision and recall for class 1 but low precision and recall for classes 2, 3, and 4. The overall accuracy was 67%, with a weighted average f1-score of 0.64. The macro average f1-score, giving equal weight to each class, was 0.53.

The RF classifier model was optimized using a grid search with parameters: max depth of 2, n estimators of 5000, and random state of 0. After applying the model to test data, predictions were saved in an Excel file for analysis. The model's accuracy on the test data were approximately 41.47%, showing its ability to predict traffic accident severity in 41.47% of cases.

Attributes	Mapping
Accident Index	
Date	
Day of week	1—Sunday, 2—Monday, 3—Tuesday, 4—Wednesday, 5—Thursday, 6—Friday, 7—Saturday
Time of Accident, Accident Location—A	1—Urban, 2—Rural, 3—Unallocated
Accident Location-A Chainage-km	
Accident Location-A Chainage-km-RoadSide	LHS, RHS
Nature of Accident—B1, B2, B3	1—Overturning, 2—Head on collision, 3—Rear End Collision, 4—Collision Brush/Side Wipe, 5—Right Turn Collision, 6—Skidding, 7a—Others-Hit Cyclist, 7b—Others-Hit Pedestrian, 7c—Others-Hit Parked Vehicle, 7d—Others-Hit Fixed Object, 7e—Others-Wrong Side Driving, 7f—Others-Hit Animal, 7g—Others-Hit Two-Wheeler, 7h—Others-Unknown, 7i—Others-Fallen down, 8—Overtaking vehicle, 9—Left Turn Collision
Accident Severity—C	1—Fatal, 2—Grevious Injury, 3—Minor Injury, 4—Non-Injury (Damage only)
Classification of Accident—C1, C2, C3	1—Fatal, 2—Grevious Injury, 3—Minor Injury, 4—Non-Injury (Damage only)
Causes—D1, D2, D3, D4, D5	1—Drunken, 2—Overspeeding, 3—Vehicle out of control, 4a—Fault of driver of motor vehicle, 4b—Driver of other vehicle, 4c—Cyclist, 4d—Pedestrian, 4e—Passenger, 4f—Animal, 5a—Defect in mechanical condition of motor vehicle, 5b—Road condition
Road Feature—E	1—Single lane, 2—Two lanes, 3—THREE lanes or more without central divider median, 4—Four lanes or more with central divider along with carriageway width
Road Condition—F	1—Straight Road, 2—Slight Curve, 3—Sharp Curve, 4—Flat Road, 5—Gentle incline, 6—Steep incline 7—Hump, 8—Dip
Intersection Type—G	1—T Junction, 2—'Y Junction, 3—Four arm junction, 4—Staggered junction, 5—Roundabout, 6—Uncontrolled junction
Weather Conditions—H	1—Fine, 2—Mist/Fog, 3—Cloud, 4—Light Rain, 5—Heavy Rain, 6—Hail/sleet, 7—Snow, 8—Strong Wind, 9—Dust Storm, 10—Very Hot, 11—Very Cold, 12—Other extraordinary weather condition
Vehicle Type Involved—J—V1, V2, V3, V4	1—Car/Jeep/Van, 2—SUV, 3—Bus, 4—Mini Bus, 5—Truck, 6—Two—Wheeler, 7—Three—Wheeler, 8—Cycle, 9—Pedestrian, 10—Tractor, 11—Unknown, 12—Animal, 13—Objects, 14—LCV, 15—MAV
Number of Vehicles	
Number of Casualties-Fatal, Grievous Injury, Minor Injury, Non Injured	

Table 1.
Dataset attributes and parameters mapping.

6.3.1 Prediction output

6.3.1.1 Comparison between observed and predicted accident severity levels

The predicted values are generated by the RF model using the input features, while the actual accident severity indices are represented by the observed values. **Figures 2** and **3** summarizes the comparison between observed and predicted values.

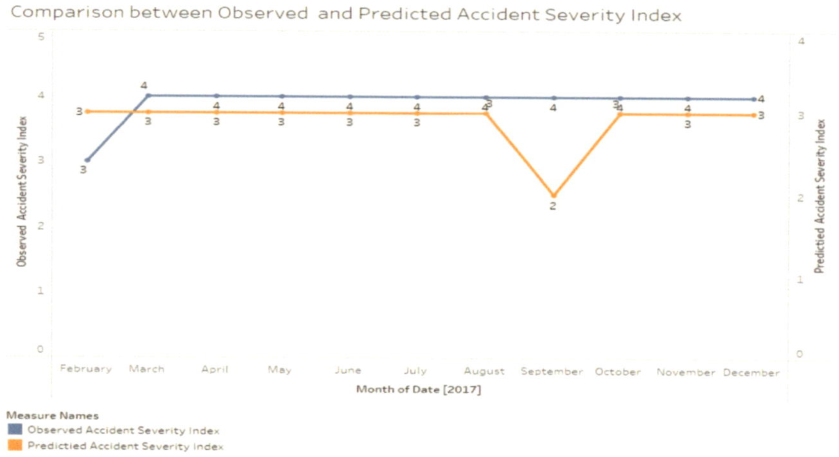

Figure 2.
Comparison of accident severity as observed and predicted index.

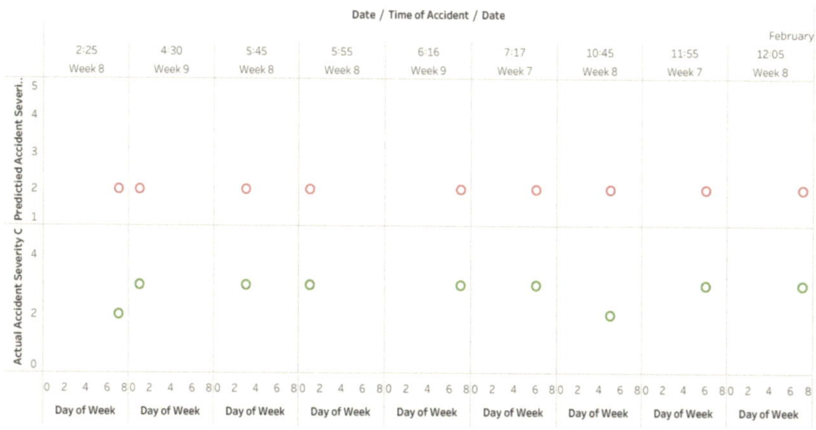

Figure 3.
Comparative analysis of observed and predicted accident severity index against time.

The RF model accurately predicts the accident severity index on dates like 25-02-2017, 17-04-2017, and 22-04-2017. On 18-02-2017, 23-02-2017, and 27-03-2017, the model predicted lower accident severity index values than observed. On 24-05-2017 and 20-10-2017, the model occasionally overestimated the accident severity index by predicting a higher value than observed.

The model may have a bias due to an imbalance in the training dataset, with severity index 2 occurring more frequently than other categories. This bias is evident when the model often predicts a severity index of 2 for accidents, even when the observed values differ.

6.3.1.2 Comparative analysis of observed and predicted accident severity index against time

The plot for the 165 rows of predicted data does not fit in the A4 sheet. **Figure 4** displays the date, day of the week, and time of the accident, as well as the observed

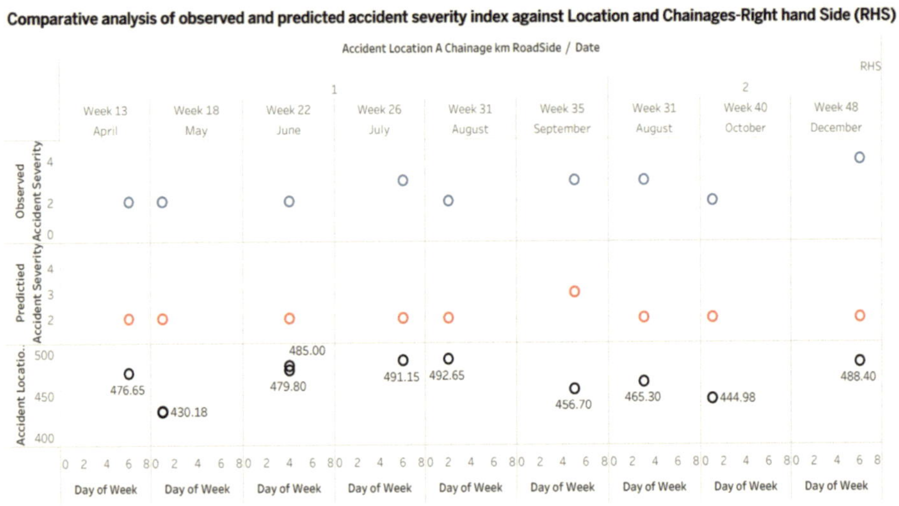

Figure 4.
Comparison between the actual and predicted severity of accidents based on location and chainages-(RHS).

and predicted accident severity indices. The data are published and the link is provided in the Tableau graphs visuals availability [A-i].

The dataset contains accident data from February 18 to December 31, 2017, analyzed using Tableau from the Excel table provided.

The accident severity index ranges from 1 to 4, with 1 being the least severe and 4 the most severe.

The majority of accidents in the dataset have a severity index of 3 or 4. A severity index of 2 indicates a less severe accident, while 4 indicates a more severe one. Most accidents are predicted to have a severity index of 2 or 1. The predicted severity index is typically lower than the observed severity index, indicating room for improvement in the accuracy of the accident severity prediction model.

6.3.1.3 Comparison between the actual and predicted severity of accidents based on location and chainages on the right-hand side (RHS)

The Tableau plot (**Figure 5**) displays accident data on the right side of the road. It shows the date, day of the week, accident location, observed accident severity index, and predicted accident severity index for each incident.

The data are published and the link to the Tableau graphs provided in the availability of the visual [A-ii]. The plot effectively shows the spatial distribution of accidents and their severity over time, helping identify patterns and trends. The Tableau plot does not fit on an A4 sheet.

The majority of accidents have an observed severity index of 2 or 3, indicating moderate severity. However, the predicted accident severity index largely remains at 2, suggesting somewhat conservative predictions that do not fully capture the observed severity range.

External factors like traffic patterns or weather conditions may have a greater impact on the occurrence and severity of accidents than the day of the week. There seems to be no correlation between the day of the week and the frequency or severity of accidents.

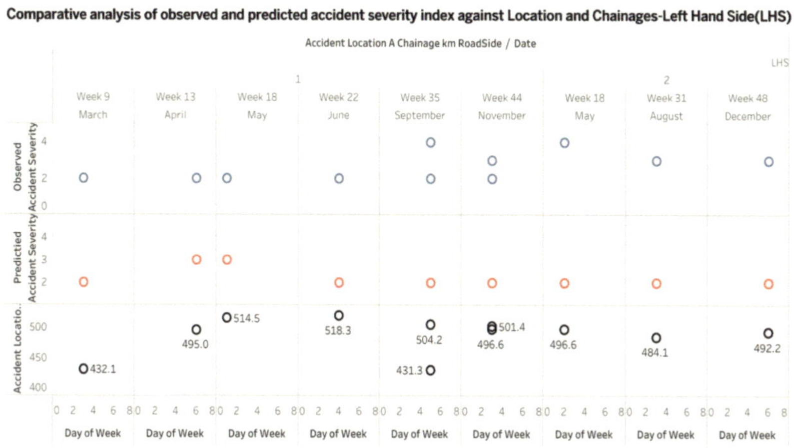

Figure 5.
Comparison between the actual and predicted severity of accidents based on location and chainages on the LHS.

6.3.1.4 Comparison between the actual and predicted severity of accidents based on location and chainages on the left-hand side (LHS)

The data are published and the link is provided in the Tableau graphs visuals availability [A-iii]. The graph displays the date, day of the week, and accident location on the Left Hand Side (LHS) of the road, along with observed and predicted accident severity indices. The plotted predicted data does not fit on an A4 sheet.

The majority of accidents on the left side of the road had a severity index of 2 or 3, indicating that most collisions were of moderate severity. Few instances of severity index 1 and 4 were observed.

The predictive model may be biased toward predicting less severe accidents, as the majority of cases had a predicted accident severity index of 2, with only a few instances of values 3 and 4.

The day of the week may not be a significant predictor of accident severity on the left side of the road, as accidents appeared to occur every day without a discernible pattern or trend.

There may not be a specific accident hotspot or concentration on the left-hand side of the road, as the accident locations were scattered along the roadway at various distances, measured by Accident Location-A Chainage km.

6.3.1.5 Data recording and availability

To enhance road safety modeling accuracy, India needs a more advanced data recording system for road accidents. This system should comply with MoRTH and IRC guidelines and utilize the Road Accident Recording and Reporting Formats. Digital monitoring can increase data collection frequency and minimize missing information. Machine learning can help regain missing data, improving road safety modeling accuracy.

6.3.2 Conclusion

To improve the model's performance, correct dataset imbalance and adjust hyperparameters. The RF classifier predicted traffic accident severity with 67% accuracy on the training set and around 41.47% on the test set. It tended to underestimate severity, possibly due to bias in the training data. No clear link was found between the day of the week and accident occurrence or severity.

No discernible patterns or trends were observed in terms of accident location. The model frequently underestimated accident severity, although it accurately predicted it in some instances. External factors may have a greater influence on the occurrence and severity of accidents. The observed and predicted accident severity indices were compared against variables such as dates, times, and locations on both sides of the road.

To enhance road safety modeling, adopting a sophisticated data recording system in line with MoRTH and IRC recommendations is crucial. Digital monitoring of road accidents can boost data collection frequency and prevent vital information loss. Incorporating machine learning techniques can improve interventions and decision-making in traffic accident prevention and mitigation.

Our research in accident severity modeling stands out for leveraging Artificial Intelligence (AI) models, specifically the Random Forest (RF) algorithm. We focus on improving accuracy and providing tailored solutions for India's road safety challenges. Our work is a standard for precise and reliable accident severity predictions with global applicability. This study contributes to the literature in this field.

6.3.3 Future scope

The study presented offers a solid foundation for future research in the area of road safety modeling and accident prevention on Indian highways. Despite the constraints of the current study, it highlights potential areas for further research, which will be explored in subsequent studies.

The study has recognized the presence of dataset bias and imbalance that could impact the performance of the model. Subsequent research will prioritize enhancing both the quality and quantity of data to mitigate bias and enhance model performance. This will entail investigating alternative data sources, refining data collection techniques, and resolving data quality concerns.

The study employed the Random Forest (RF) algorithm to construct a predictive model for the severity of traffic accidents. Future research will investigate the utilization of alternative machine learning algorithms or ensemble models to enhance the model's performance. Furthermore, efforts will be made to refine hyperparameters and rectify dataset imbalance to enhance the accuracy of the model.

The analysis of external factors in accidents was emphasized in the study, focusing on their influence on predicting accident severity. Future research should investigate the effects of external factors like weather conditions, road infrastructure, and driver behavior on accident severity. This research can improve the precision of predictive models and provide valuable insights for decision-making in accident prevention strategies.

The study emphasized the necessity of implementing an advanced data recording system that complies with the guidelines established by MoRTH and IRC. Subsequent research could concentrate on the creation of a real-time monitoring system capable of collecting road safety data instantly and offering valuable information for initiatives aimed at preventing accidents.

7. Integrating machine learning in road safety applications: a paradigm shift toward taking preventative measures

The use of machine learning in road safety applications represents a fundamental shift from reactive approaches to preventing high-risk incidents through data-driven interventions. Such smart systems that can predict and prevent road accidents using machine learning on large datasets aid save lives as well as promote safer transport systems in general.

Use of predictive analytics in preventing accidents. To road safety, the prediction aspect plays a vital role in its application in this field due to its ability to predict the likelihood of events. ML algorithms, can, therefore, use historical accident records together with traffic patterns and other related information contained in the environment to predict where over time such events will be more likely to happen thus preventive measures should be put in place. One of the interventions implemented is Predictive Policing, which involves the deployment of law enforcement officers, either on foot or in vehicles, to specific high-risk areas where accidents are most common. This deployment occurs during peak hours each day, extending late into the night, with the exception of major routes leading out of town. Dynamic Traffic Management consists of promptly modifying speed limits, lane arrangements, or signal timing based on the current traffic conditions, such as congestion areas on highways. On the other hand, Customized Driver Notifications involve sending personalized alerts to drivers, taking into consideration their location, and driving mode, among other factors, to inform them about potential hazards.

The design and maintenance of road infrastructure can be improved using machine learning methods. This will help to promote driving safety since it becomes possible to have a more comprehensive examination of accident data alongside other aspects like pavement conditions, road geometry, and traffic flow patterns. Dangerous spots on roads need to be defined. These spots are usually more prone to accidents caused by poor road design, lack of proper maintenance, or insufficient road signs. Road design optimization involves the provision of guidelines on road geometry, road safety features implementation, and measures enhancing visibility in order to prevent traffic accidents. For this reason, there is an emphasis on maintenance activities such that road maintenance agencies can proactively address road defects and associated infrastructural failures leading to accidents that could have been avoided. The use of machine learning (ML) enabled Advanced Driver Assistance Systems (ADAS) is one such driving safety supplement since they form protective equipment inside the drivers' car. The technology in the use of such systems entails sensors, cameras, and advanced algorithms that include the function called Lane Departure Warning for example. It alerts drivers when their vehicles move out of designated lanes hence averting unintended lane departures There is an Adaptive Cruise Control (ACC) that enables cars to vary their speed maintaining them at a safe distance from cars in front brake quickly enough to accommodate those behind in case of sudden stops thereby minimizing chances of rear-end collisions. The term Automatic Emergency Braking (AEB) is used to describe technology that detects potential accidents and then applies brakes to prevent or lessen their impact customized driver feedback and education are essential in addressing road safety concerns especially now that road carnages as a result of careless driving or mechanical problems which have become so rampant. The list includes fog, rain, and ice as well as construction zones or animal crossings as some of the factors to be considered. We can achieve more personalized driver feedback and an enhanced driving experience utilizing machine learning which aids individuals in adopting safe driving habits. It uses data obtained from smartphones, within-car sensors, and telematic devices to measure the behavior of drivers during the most dangerous sections of the road Additionally, this allows the detection of risky driving patterns seen as overspeeding, abrupt stops, and driving distractedly. Providing personalized feedback helps convince individuals to shun dangerous practices of driving for safer ones. Gamification has become inevitable to enhance road safety by incentivizing those drivers committer safe actions at all times Factors to consider Ensuring the safety of roads through machine learning involves taking different factors into account. How machine learning performs in such scenarios greatly depends on data quality and quantity. Hence in these cases, the appropriate action that is taken ensures that the data used for learning is accurate, comprehensive, and unbiased. One of the major concerns that might crop up when dealing with machine learning involves biases and fairness. When this happens, these issues could lead to unjust or discriminatory outcomes toward marginalized communities due to the replication of training data biases by models. As such, addressing issues concerning model fairness and bias should be a major concern in solving these problems Concerns about privacy may arise from collecting and analyzing huge amounts of data hence many people tend to be concerned when they are dealing with data in any activity. Transparency and ethics should be observed while handling this matter. To sum up, there is still great potential for this kind of technology to help us reduce road accidents at a very significant level while improving road safety through machine learning. The primary focus of this study is to investigate how machine learning algorithms can be applied to improve road safety. Machine learning can cause a revolution in road safety projects. In this particular context if data are proactively used instead of waiting for accidents they can have far fewer preliminary fatalities and this way move to a safer public transportation system. Thus, as already said above there are various advantages that come with the following methodical approach: Predictive analytics can be a useful tool for identifying possible risks beforehand. For example, some of these risks include accidents that take place in rainy conditions where high risks are involved. Within this field, among others, we have predictive policing, dynamic traffic control, and personalized driver notifications. It is essential to have a look at crashes along with infrastructure data by use of machine learning algorithms that enable identification of dangerous spots on roads for improvement in design hence optimization of maintenance activities which results into much safer roads. This contributes to the enhancement of road safety. These vehicle systems that rely on machine learning for instance the advanced driver assistance systems (ADAS) greatly enhance driver consciousness as they improve road safety They offer functions such as adaptive cruise control, autonomous emergency braking, and lane departure warning that work for all drivers regardless of sex or age while on the road. The study aims at improving road safety by the best means available which is machine learning techniques.

8. Conclusions

Improving safety on Indian roads by using predictive analysis and machine learning algorithms for accident severity modeling seems a promising avenue for addressing the crucial issue of road traffic accidents. In this regard, the focus of the study is on the possibility of using data analytics coupled with advanced analytics to come up with better strategies for preventing accidents and improving safety.

In predicting accident severity and spotting risky zones, Random Forest (RF) classifiers in particular have potentials that are enormously great when it comes to exploiting machine learning approaches. For example, among the various challenges faced lie imbalanced datasets as well as poor quality data collection methods but such problems are manageable based on this study indicating how they can be used for providing direction toward interventions meant to enhance road safety.

The research suggests that provided there is already enough database structure as well as continuous machine learning models' adjustment one can enhance prediction about accident severity with respect to its accuracy and consistency. Such can also form input in policy policy-making process, resource allocations, or specifically into designating safer roads accordingly while taking into account various factors among which include types of roads, crossroads, etc. trailed information outlined by this research.

Integrating machine learning in strong data collection and analysis initiatives can enable stakeholders to transition from a reactive approach to managing traffic safety into a predictive one. This change in focus will help minimize deaths and financial losses from accidents on the roads while at the same time reducing the occurrence rate as well as the seriousness of accidents recorded annually.

9. Data availability, software availability, tableau graphs visual availability

Ref. [15].

9.1 Data availability

The Data of Accident Severity Prediction Modeling for Indian Highways Case Study stretches mentioned is available on Zenodo Open Access Repository and available for further analysis at https://doi.org/10.5281/zenodo.7773156 [16].

Data are available under the terms of the Creative Commons Attribution 4.0 International license (CC-BY 4.0).

9.2 Software availability

https://github.com/humera-k/RF_Accident_Severity

https://zenodo.org/badge/latestdoi/616376786

Data are available under the terms of the Creative Commons Attribution 4.0 International license (CC-BY 4.0).

9.3 Tableau graphs visual availability

Ref. [16].

https://public.tableau.com/app/profile/humera.khanum/viz/Accidental_Analysis_1/Sheet52 (Comparative analysis of observed and predicted accident severity index against time)

https://public.tableau.com/app/profile/humera.khanum/viz/Accidental_Analysis_1/Sheet3 (Comparative analysis of observed and predicted accident severity index against Location and Chainages-Right hand Side (RHS))

https://public.tableau.com/app/profile/humera.khanum/viz/Accidental_Analysis_1/Sheet4 (Comparative analysis of observed and predicted accident severity index against Location and Chainages-Left Hand Side (LHS))

Acknowledgements

Our sincere thanks are extended to the National Highways Authority of India and ILFS Engineering and Construction Company for their invaluable assistance in providing us with raw data on accidents. Their crucial support has played a key role in facilitating the execution of this research and analysis.

Conflict of interest

The authors declare no conflict of interest.

Abbreviations

RF	random forest
MoRTH	Ministry of Road Transport and Highways
NHAI	National Highways Authority of India

Author details

Humera Khanum[1,2*†], Rushikesh Kulkarni[1†], Anshul Garg[2†] and Mir Iqbal Faheem[3]

1 Civil Engineering Department, Symbiosis Institute of Technology, Pune Campus, Symbiosis International (Deemed University), Pune, India

2 School of Civil Engineering, Lovely Professional University, Phagwara, Punjab, India

3 Civil Engineering Department, Deccan College of Engineering and Technology, Hyderabad, India

*Address all correspondence to: author1@inbox.com

† These authors contributed equally.

References

[1] Global Status Report on Road Safety 2018. Geneva: World Health Organization; 2018

[2] Patel M, Patel R. A study on causes of road accidents in India. International Journal of Engineering Research and Applications. 2013;**3**(6):1386-1391

[3] Yan M, Shen Y. Traffic accident severity prediction based on random forest. Sustainability (Switzerland). 2022;**14**(3):2. DOI: 10.3390/su14031729

[4] Wang C, Quddus MA, Ison SG. The effect of traffic and road characteristics on road safety: A review and future research direction. Safety Science. 2013. ISSN: 09257535;**57**:264-275. DOI: 10.1016/j.ssci.2013.02.012

[5] Barbosa P, Andrade M, Ferreira S. Machine learning applied to road safety modeling: A systematic literature review. Journal of Traffic and Transportation Engineering (English Edition). 2020;**7**(6):775-790. DOI: 10.1016/j.jtte.2020.07.004

[6] Al-Mistarehi BW, Alomari AH, Imam R, et al. Using machine learning models to forecast severity level of traffic crashes by R studio and ArcGIS. Frontiers in Built Environment. 2022;**8**:1-14. DOI: 10.3389/fbuil.2022.860805

[7] Lord D, Mannering F. The statistical analysis of crash-frequency data: A review and assessment of methodological alternatives. Transportation Research Part A: Policy and Practice. 2010;**44**:291-305. DOI: 10.1016/j.tra.2010.02.001

[8] Ramanujam V, Bhalla K. Speeding on Indian roads: A survey of Indian drivers. Accident Analysis and Prevention. 2009;**41**(3):527-532. DOI: 10.1016/j.aap.2009.01.009

[9] Daniel MC, Woo KH. Risky behaviors and road safety: An exploration of age and gender influences on road accident rates. PLoS One. 2024;**19**(1):e0296663. DOI: 10.1371/journal.pone.0296663

[10] Hoang Long V, Ahmed K, Ma W. A random forest approach to predicting traffic accident severity. IEEE Access. 2021;**9**:1219-1232. DOI: 10.1109/ACCESS.2021.3098040

[11] Singh G, Kumar A. Random forest-based prediction model for traffic accident severity on Indian highways. Journal of Traffic and Transportation Engineering (English Edition). 2021;**8**(6):693-706. DOI: 10.1016/j.jtte.2021.05.012

[12] Road Safety Manual a Guide for Practitioners: Road Safefy Management. PIARC. Version 1-20/10/2015; 2019. p. 36. Available from: https://roadsafety.piarc.org/en/road-safety-management

[13] Damodariya SM, Patel CR. Identification of factors causing risky driving behavior on high-speed multi-lane highways in India through principal component analysis. International Journal of Engineering. 2022;**35**(11):2130-2138

[14] Santos D, Saias J, Quaresma P, Nogueira VB. Machine learning approaches to traffic accident analysis and hotspot prediction. Computers. 2021;**10**(12):157

[15] Humera K, Anshul G, Iqbal FM. Accident severity prediction modeling for road safety using random forest algorithm: An analysis of Indian

highways. F1000Research. 2023;**12**:494.
DOI: 10.12688/f1000research.133594.2

[16] Khanum H, Garg A, Faheem MI.
Data for Accident Severity Prediction
Modelling for Indian Highways Case
Study (Accidentdata_V1). Zenodo; 2023.
DOI: 10.5281/zenodo.7773156

Chapter 11

Road Maps and Sensor Integration for the Enhancement of Lane-Keeping Assistants

Emerson Pereira Cavalheri and Marcelo Carvalho dos Santos

Abstract

Current efforts of vehicle manufacturers and research groups in designing and developing safer Intelligent Transportation Systems have revolved around achieving higher levels of driving automation for on-road vehicles. However, current approaches remain unable to assure safe vehicle autonomy in all conditions. Leveraging the communication between the infrastructure, for instance, the road geometry from high-definition maps, and vehicles could be a key enabler of safer Intelligent Transportation Systems. This combination would increase the overall traffic awareness which could benefit current automation approaches. In this study, a new lane-keeping system integrating information from a road map, satellite receiver, and inertial sensors is presented. Tests driving in complex urban environments showed that the proposed system kept the vehicle centered in the lanes during long satellite outages. This result was accomplished with a novel integration between the inertial and road map where the inertial was calibrated by the Map. The position cross-track accuracy upper and lower bounds, at 95% confidence, were 3 and 1 cm from achieving the control limit level (0.1 m) for Intelligent Transportation Location Based Systems. With these results, this work provides a new contribution to increase the robustness of current lane-keeping assistant approaches.

Keywords: lane-keeping assistant systems, road maps, sensor fusion, autonomous vehicles, sensor navigation

1. Introduction

Road traffic deaths reach 1.19 million every year, being the leading cause of death for people aged between 5 and 29 years old (WHO) [1]. To reduce road accidents, several efforts in the development of vehicle safety systems, Advanced Driver Assistance Systems (ADAS), and Autonomous Vehicles (AV) technologies have been made.

A common approach to autonomous driving is to use machine learning techniques to learn the driving behavior from camera and other sensor data. This approach is restricted to the training data variability (light, weather, etc...) and location, thus being restricted in its Operational Design Domain (ODD) [2].

While higher levels of autonomy are still a few years ahead in the future, reliable Lane Departure Warning (LDW) and Lane-Keeping Assistant Systems (LKAS) could prevent collisions [3]. Current LKS have been mostly done using imagery sensors, such as cameras, to identify the lanes. Several road boundary detection and following have been developed assuming the road scene is identifiable [4]. However, combined with the road environment complexity, imagery systems performance for lane detection is affected by several factors: dependence of natural light, alternation of structured and unstructured roads; and appearance of roads in weather changes and scenes with shades and darkness [5, 6]. Two fatal autopilot incidents have demonstrated the limitation of cameras in poor light conditions and this alerted the industry and community that current systems still need major improvements [7, 8].

The accurate tracking of the vehicle's position is an essential task to safely keep the vehicle in the correct lane of navigation. This task, also known as vehicle localization or positioning, is equipped with several sensors to estimate the vehicle position, velocity, and attitude (PVA) in space using a sensor fusion algorithm, e.g. the Kalman filter (KF). A commonly used approach in vehicle localization algorithms is the fusion of Global Navigation Satellite Systems (GNSS) with Inertial Navigation Systems (INS). GNSS errors are relatively noisy from the second to second period and have long-term accuracy, whereas INS exhibits relatively short-term low noise errors but tends to drift over time [9]. This combination has a complementary nature and the potential to provide accurate solutions anywhere on the Earth's surface. However, during long GNSS outages, INS alone does not offer a reliable solution due to its inherent errors. Thus, additional information to provide a continuous and reliable solution needs to be explored along with different integration approaches.

Another piece of information that could provide extra robustness to vehicle navigation sensor fusion, which is neither affected by the surrounding environment nor by sensor noise, is the use of a High Definition (HD) map of the roads. For instance, applications such as curve speed warning, vision enhancement, path prediction, LKAS, and collision warning can all be enhanced with the use of an HD map of the roads and, consequently, provide large-scale autonomous navigation [10]. Some applications, integrating sensor fusion and maps along with a map-matching algorithm, have shown improvements in solution availability [11, 12]. Other contributions using a 3-D map of cities could predict non-line-of-sight (NLOS) satellite signals and improve 2-D positioning in dense urban areas [13, 14]. In the above-mentioned contributions, the map information and sensor fusion were used in separate filters, and the map-matching algorithm would connect them in a further step of the algorithm. It was also observed an improvement in reliability with low positioning accuracy. Moreover, the map-matching approaches in the literature were dependent on the correct road identification step, which can generate ambiguous matches. Thus, applications using road maps can be further improved; for instance, by inputting the map space directly into the navigation filter, navigation constraints, and calibration of sensors could be performed.

In summary, there are many challenges to providing a safe and reliable lane-keeping vehicle navigation system. Among these challenges, LKAS at a global scale must be proposed to enable AV systems without restrictions. To achieve this, the accurate track of the vehicle position on its current lane of navigation operating in any environment is the major task to be solved. This way, a road map could provide extra information to GNSS/INS sensor fusion and support its current limitations. With the intent of providing a generic approach for LKAS, we propose a novel approach combining map of the lanes information with current sensor fusion techniques using satellite and inertial information. This map of lanes is obtained before navigation with any mapping techniques, and it precisely represents the location of the center of lanes, which can be retrieved during navigation. This GNSS/INS with map fusion can track the vehicle position at all times and potentially provide a global and scalable solution. The objective of this system is to constantly track the vehicle's position on the road at all times and in any environment. Another contribution of this system is to provide optional redundant information to enhance the safety of current LKAS.

2. Sensor navigation overview

Vehicle navigation can be represented by positions, velocities, and orientation states. Common sensors used to track either directly or indirectly are Inertial Measurement Units (IMUs), Global Navigation Satellite Systems (GNSS), wheel odometers, cameras, Light Detection And Ranging (LiDARs), RAdio Detection and Ranging (RADARs), and others. This work will use IMUs and GNSS as sensors; thus a brief explanation is given in Sections 2.1 and 2.2. To convert sensor measurements into parameters of interest, and track them over time, pre-processing and filtering are required. In 2.4, the pre-processing of IMU data and GNSS, and INS fusion techniques are described.

2.1 IMU navigation

IMUs sense rotations and accelerations in three orthogonal axes. The combination of these two sensors is capable of producing a navigation solution independent of any other sensor. This solution, also known as the Inertial Navigation System (INS), can be obtained by integrating the angular rates and specific force over time. The solution is composed of position, $\boldsymbol{r}_{pb}$, velocity, $\boldsymbol{v}^b_{ib}$, and attitude, $\boldsymbol{\psi}^b_{ib}$, vectors, also referred to as a PVA solution. The IMU measurements are affected by errors and must be corrected before computing a PVA solution. Depending on the quality of the IMU, these errors greatly impact the computation of PVA, in some cases, making IMU navigation by itself impracticable. Details on IMU quality, error sources, and how to mitigate them see [15].

2.2 GNSS

GNSS solution is composed of accurate 3-D position, velocity, and timing by transmitting radio signals from satellite constellations and the user's receiver antenna. This solution is obtained by computing ranges from measured time or phase difference between received signals and receiver-generated signals. The transmitted messages have ranging code signals, navigation data stream with satellite orbits, and clock corrections to GNSS system time. GNSS observables can be computed using the transmitted message contents.

The basic observables are pseudoranges, carrier-phase, and Doppler. Examples of fully operational constellations including the United States Global Positioning System (GPS) and Russian Global'naya Navigatsionnaya Sputnikova Sistema (GLONASS).

There are several GNSS positioning modes that vary in terms of processing, modeling, physical structure, and precision and accuracy. The most common algorithm used in every GNSS positioning filter is the single point positioning (SPP). The most accurate navigation solutions are achieved using the relative mode. Another well-established processing mode, due to its simplicity in physical structure but complexity in modeling, is precise point positioning (PPP). PPP uses a single receiver to obtain centimeter-to-decimeter accurate positions. PPP started mainly as a post-processing technique; however, advances in real-time corrections are allowing PPP in real-time. An in-depth discussion of GNSS processing modes can be found in Refs. [16–18]. In this work, a PPP processing mode is utilized.

2.3 Road maps

Since the advances of mapping techniques in the 90s, there has been an increasing investment of companies in producing maps for ADAS [19, 20]. The main reason is the wide range of applications and impacts on safety systems maps can provide, for instance: curve speed warning, vision enhancement, path prediction, lane-keeping assistance, collision warning, and especially the development of autonomous navigation systems [10, 21]. Examples of mapping products are Mobileye's Road Experience Management) and TomTom's RoadDNA. RoadDNA is generated with the combination of mobile mapping using LiDARS and other sensors. It also contains lane information such as width, height, markings, and speed limits. By providing road-way profile, curvature, and terrain information, these maps can extend the user's view for better vehicle prediction. HD maps are key information to enable fully autonomous driving. An example of the RoadDNA is shown in **Figure 1**.

2.4 Filters

2.4.1 Kalman filter

The task of tracking the state changes over time is accomplished by the Kalman filter (KF) using the sensor data. KF is composed of two main steps, the system propagation and measurement update. Before system propagation, state and error covariance are represented as: $\hat{\boldsymbol{x}}_k^-$ and $\boldsymbol{P}_k^-$, respectively. After the measurement

Figure 1.
HD RoadDNA with 10 cm relative accuracy, from [22].

update, the notation adopted to represent the state and error covariance are: $\hat{\boldsymbol{x}}_k^+$ and $\boldsymbol{P}_k^+$, respectively. Where k represents the current epoch of update.

The system model describes how the Kalman filter states and covariance matrix vary with time, in other words, it represents the state dynamics, $\dot{\boldsymbol{x}}$, as shown in the dynamic model, in Eq. (1).

$$\dot{\boldsymbol{x}}(t) = \boldsymbol{F}(t)\boldsymbol{x}(t) + \boldsymbol{G}(t)\boldsymbol{v}_s(t), \tag{1}$$

where $\boldsymbol{F}(t)$ is a continuous function of the dynamic behavior of the states, called the system matrix, and $\boldsymbol{G}(t)$ is the continuous system noise distribution matrix. The continuous system noise vector, $\boldsymbol{v}_s(t)$, is assumed to be zero-mean Gaussian distribution with a system noise covariance:

$$\boldsymbol{Q} = \mathbf{E}\left(\boldsymbol{v}_s\boldsymbol{v}_s^T\right). \tag{2}$$

The measurement vector, $\boldsymbol{z}$, corresponds to observations of the system, described by the state-vector, made from information coming from sensors, such as position, orientation, distance ranging, and rate-rotation measurements. The measurement model is composed of the measurement matrix, $\boldsymbol{H}(t)$, which represents the relationship between states, $\boldsymbol{x}(t)$, and measurements, $\boldsymbol{z}(t)$, and a measurement white noise vector, $\boldsymbol{w}_m(t)$:

$$\boldsymbol{z}(t) = \boldsymbol{H}(t)\boldsymbol{x}(t) + \boldsymbol{w}_m(t). \tag{3}$$

An important quantity during the measurement update is the *measurement innovation*, $\boldsymbol{\delta z}^-$, which represents the difference between the measurements and the modeled measurements using the propagated state vector:

$$\boldsymbol{\delta z}^- = \boldsymbol{z} - \mathbf{h}(\hat{\boldsymbol{x}}^-). \tag{4}$$

The quality of the measurements is given by the measurement covariance matrix, $\boldsymbol{R}$. Defined as the expectation of the square of the measurement noise, KF assumes to be Gaussian and uncorrelated in time:

$$\boldsymbol{R} = \mathbf{E}\left(\boldsymbol{w}_m\boldsymbol{w}_m^T\right). \tag{5}$$

The measurement covariance matrix values come from the sensors' measurement properties.

In Appendix A, the discrete KF, describing a sequence of steps to run the KF estimation, is presented.

2.4.2 GNSS/INS integration

An important aspect of GNSS/INS integration is the complementary nature of these systems. GNSS has global coverage and is accurate over long periods, whereas INS is a self-contained sensor with high accuracy over short periods. This way, GNSS can calibrate the INS drift, and INS can bridge GNSS solution during outages.

Several integration architectures combine GNSS and INS measurements [9]. If the integrated PVA solution is acquired by combining GNSS and INS raw measurements, a Tightly Coupled (TC) integration strategy is adopted. When GNSS and INS PVA solutions are combined, the integration strategy is called Loosely Coupled (LC). These integration strategies are described in 2.4.4 and 2.4.3, respectively.

In the GNSS/INS error-state integration, the final solution is the corrected INS navigation, called closed-loop correction. Thus, the final PVA solution is obtained with the following equations:

$$\hat{C}^e_b(+) = \delta\left(\hat{C}^e_b\right)^T \hat{C}^e_b(-) \approx \left(I_3 - \left[\delta\hat{\Psi}^b_{ib}\wedge\right]\right)\hat{C}^e_b(-), \tag{6}$$

$$\hat{v}^e_{eb}(+) = \hat{v}^e_{eb}(-) - \delta\hat{v}^e_{eb}, \tag{7}$$

$$\hat{r}^e_{eb}(+) = \hat{r}^e_{eb}(-) - \delta\hat{r}^e_{eb}, \tag{8}$$

where $[\wedge]$ denotes the skew-symmetric matrix operator of the Euler angles ($\left[\delta\hat{\Psi}^b_{ib}\wedge\right]$). $\hat{C}^e_b$ represents a rotation matrix from body (b) to Earth frames (e).

The error-state INS/GNSS integration architecture normally estimates three attitude, three velocity, and three-position errors, and six IMU accelerometer and gyroscope biases. These are the parameters for an LC approach. For a TC approach, GNSS receiver clock, δdt_r, and drift, $\delta \dot{d}t_r$, are also estimated.

The INS propagation and LC error-state vector, with 15 parameters, is given by:

$$x^e_{INS} = \begin{pmatrix} \delta\Psi^e \\ \delta v^e \\ \delta r^e \\ b^e_a \\ b^e_g \end{pmatrix}. \tag{9}$$

For a GNSS/INS TC approach, the error-state vector is composed of the following 17 parameters:

$$x^e_{GNSS} = \begin{pmatrix} x^e_{INS} \\ \delta dt_r \\ \delta \dot{d}t_r \end{pmatrix}. \tag{10}$$

The INS and GNSS systems models do not interact during the system propagation step:

$$F = \begin{pmatrix} F_{INS} & 0 \\ 0 & F_{GNSS} \end{pmatrix}, \Phi = \begin{pmatrix} F_{INS} & 0 \\ 0 & F_{GNSS} \end{pmatrix}, Q = \begin{pmatrix} Q_{INS} & 0 \\ 0 & Q_{GNSS} \end{pmatrix}, \tag{11}$$

with state vector:

$$x = \begin{pmatrix} x_{INS} \\ x_{GNSS} \end{pmatrix}. \tag{12}$$

To obtain the system propagation equations the system model, based on the parameters of the state vector 9, a time derivative of each state must be calculated. The system propagation equations derivative is presented in [15].

2.4.3 LC GNSS/INS data fusion

An LC GNSS/INS architecture integrates GNSS position and velocity and INS PVA solutions in the filter to estimate a full PVA solution. The accelerometer and gyroscopes biases are fed back to correct INS navigation. A loosely coupled architecture is shown in the flowchart of **Figure 2**. The dashed lines represent closed-loop computations.

Also known as *position-domain* filter, the main benefits of this architecture are simplicity and the number of parameters are fixed. The main drawbacks are the necessity of having two filters, the GNSS and INS/GNSS, thus increasing processing time, and a minimum of four satellites must be observed to obtain a solution, otherwise the data tracked of three or less satellites are not utilized.

In the LC architecture, the measurements are composed by the GNSS and INS propagated position and velocity solutions. Thus, the innovation vector is:

$$\delta \boldsymbol{z}_k^e = \begin{pmatrix} \hat{\boldsymbol{r}}^e_{eG} - \hat{\boldsymbol{r}}^e_{eb} \\ \hat{\boldsymbol{v}}^e_{eG} - \hat{\boldsymbol{v}}^e_{eb} \end{pmatrix}_k, \tag{13}$$

where the subscript G denotes GNSS.

The measurement matrix is given by taking the first derivative of the measurement model 3 with respect to the state vector in 9:

$$\boldsymbol{H}_k^e = \left.\frac{\partial \mathbf{h}(\boldsymbol{x}, t_k)}{\partial \boldsymbol{x}}\right|_{\boldsymbol{x}=\hat{\boldsymbol{x}}_k^-} = \left.\frac{\partial \boldsymbol{z}(\boldsymbol{x}, t_k)}{\partial \boldsymbol{x}}\right|_{\boldsymbol{x}=\hat{\boldsymbol{x}}_k^-}. \tag{14}$$

Thus, by neglecting lever arm effects, the LC measurement matrix can be approximated to

$$\boldsymbol{H}_k^e \approx \begin{pmatrix} \mathbf{0}_3 & \mathbf{0}_3 & -\boldsymbol{I}_3 & \mathbf{0}_3 & \mathbf{0}_3 \\ \mathbf{0}_3 & -\boldsymbol{I}_3 & \mathbf{0}_3 & \mathbf{0}_3 & \mathbf{0}_3 \end{pmatrix}_k. \tag{15}$$

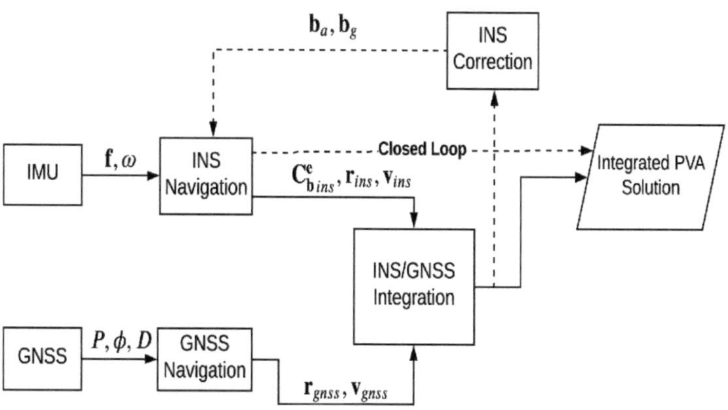

Figure 2.
Loosely coupled GNSS/INS integration.

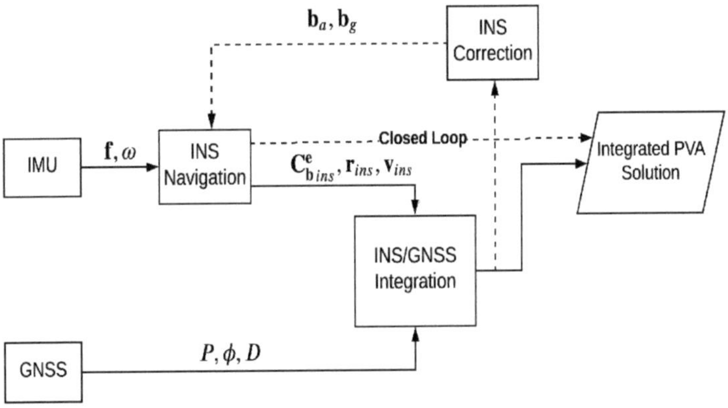

Figure 3.
Tightly coupled GNSS/INS integration.

2.4.4 TC GNSS/INS data fusion

A TC integration is a *measurement-domain* filter. It uses the GNSS raw measurements as pseudorange, carrier-phase, and Doppler shift and INS PVA solutions to the filter instead of pre-processed position and velocity solutions, as in the loosely coupled architecture. Similar to the loosely coupled, the tightly coupled feedbacks corrections to compensate accelerometer and gyroscopes measurements. In **Figure 3**, a TC GNSS/INS integration architecture is presented.

The first obvious advantage of the tightly coupled is the use of a single filter. Other advantages include: the INS errors are better estimated due to the GNSS covariance variation with satellite geometry, and even when less than four satellites are tracked, the satellite data can be used to aid the INS in computing an integrated solution [15]. The main limitation over the loosely coupled integration is the increased size of the number of parameters, which increases the computational efforts [23].

In TC architecture, the measurements are the GNSS pseudorange and carrier-phase observables. The TC measurement innovation is written by the difference between GNSS pseudorange, $\tilde{P}$, and carrier-phase, $\tilde{L}$, measurements and the computed pseudorange, $\hat{P}$, and carrier-phase, $\hat{L}$, values are derived from the current INS navigation PVA solution, per m-satellites available. Thus:

$$\delta z^e_k = \begin{pmatrix} z^e_P \\ z^e_L \end{pmatrix}_k, \quad \begin{matrix} \delta z^e_P = \left(\tilde{P}^1_G - \hat{P}^1_{INS}, \tilde{P}^2_G - \hat{P}^2_{INS}, \cdots, \tilde{P}^m_G - \hat{P}^m_{INS}\right)_k \\ \delta z^e_L = \left(\tilde{L}^1_G - \hat{L}^1_{INS}, \tilde{L}^2_G - \hat{L}^2_{INS}, \cdots, \tilde{L}^m_G - \hat{L}^m_{INS}\right)_k \end{matrix}, \tag{16}$$

where the subscript G denotes GNSS, $\hat{P}$ and $\hat{L}$ are described in 2.4.4 and 2.4.3 respectively.

The TC measurement matrix is given by 14 with respect to the state vector parameters in 10, thus:

$$\boldsymbol{H}_k^e = \begin{pmatrix} \frac{\partial \delta \boldsymbol{z}_P^e}{\partial \delta \boldsymbol{\Psi}^e} & \mathbf{0}_{m,3} & \frac{\partial \delta \boldsymbol{z}_P^e}{\partial \delta \mathbf{r}^e} & \mathbf{0}_{m,3} & \mathbf{0}_{m,3} & \frac{\partial \delta \boldsymbol{z}_P^e}{\partial \delta dt_r} & \mathbf{0}_{m,1} \\ \frac{\partial \delta \boldsymbol{z}_L^e}{\partial \delta \boldsymbol{\Psi}^e} & \frac{\partial \delta \boldsymbol{z}_L^e}{\partial \delta \boldsymbol{v}^e} & \frac{\partial \delta \boldsymbol{z}_L^e}{\partial \delta \boldsymbol{r}^e} & \mathbf{0}_{m,3} & \frac{\partial \delta \boldsymbol{z}_L^e}{\partial \boldsymbol{b}_g^e} & \mathbf{0}_{m,1} & \frac{\partial \delta \boldsymbol{z}_L^e}{\partial \delta \dot{dt}_r} \end{pmatrix}_k . \tag{17}$$

2.5 Map-matching

Road map data are interfaced with vehicle navigation applications through the use of map-matching techniques. Map-matching is often described and used in the literature as a tool to correctly identify the road segment that a vehicle is traveling on and its position on that segment [24]. The input data required to perform a map-matching are simply the position of the vehicle and a map of the roads. The process of map-matching is normally divided in two tasks: first, road segment identification, and second, vehicle positioning with respect to the selected road segment.

Several techniques are used to solve the map-matching problem, from simple geometric, topological, and probabilistic techniques to advanced Kalman filtering, fuzzy logic, and belief theory. According to Ref. [25], the identification task is the main limitation of map-matching performances in the following situations: road junctions, intersections, and complex urban areas. Another major source of possible degradation in the quality of map-matching is in the input data: the position and the map database.

Many applications use map-matching to identify the correct road of navigation and determine the vehicle location on the same segment. Another common application is to visualize the trajectory onto the road.

3. Lane-keeping system architecture proposal

The LKAS system proposed has as inputs satellite and IMU measurements with a map of lane information. The architecture of the system, referred to as (TC + LC/MAP), is composed of two data fusion approaches: a tightly coupled GNSS/INS with loosely-coupled INS/MAP. The flowchart in **Figure 4** gives a high-level view of the system architecture.

The blue boxes in **Figure 4** show the type of available data, and the red boxes represent the type of data fusion used. The "INS Navigation" and "SPP" boxes are described in 2.1 and 2.2, respectively. At the center of the system flowchart, a condition checks for satellite observation availability; if the condition is met, a TC GNSS/INS is performed; otherwise, an LC INS/MAP solution is computed to bridge the GNSS outage. The "INS correction" box in **Figure 4** corresponds to the corrected INS solution.

In the next sections, an improved map-matching (the "MM" box) process is described in 3.1, along with an explanation of the candidate search space and best candidate selection in 3.1.1 and 3.1.2, respectively. In 3.2, a description of the LC INS/MAP filter; and how the full-system solution transition is detailed, respectively. Finally, in 3.3, a position accuracy criteria based on the vehicle and road relationship is proposed.

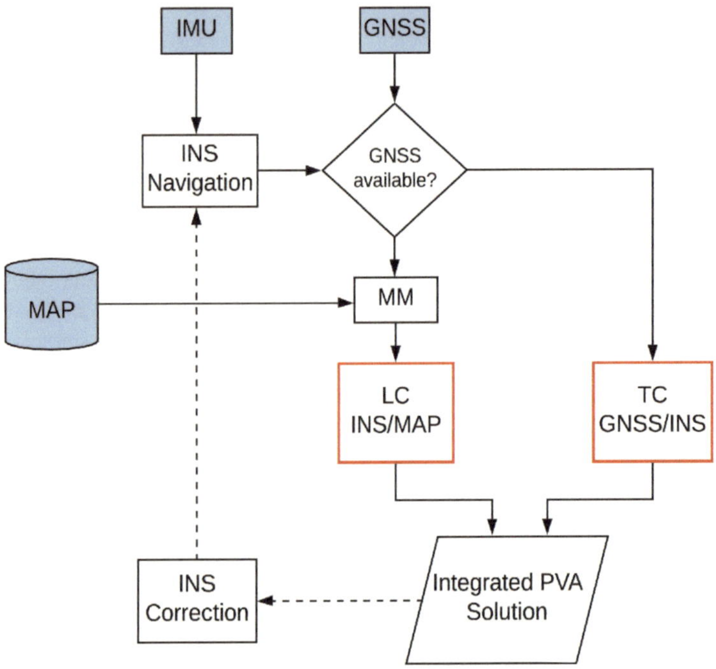

Figure 4.
Full-system architecture.

3.1 Map-matching

Standard map-matching applications start the identification process by searching for the road segment the vehicle is within a road network. If the GNSS position is degraded, ambiguous match-matching can happen. However, in this work, a fundamental assumption will be made to eliminate the problem of road ambiguity matching.

From the driver's perspective, the driving task can be described into the following: the travel plan, composed of the desired final location and vehicle starting position. With these two locations, a unique route passing through known streets can be defined. Assuming that the trajectory is fixed and no detours or overtaking other vehicles is necessary, driving becomes a constrained problem, and road identification can be eliminated with this assumption.

Then, the only task left in this map-matching is to estimate the vehicle position in the lane map trajectory. Instead of only snapping the GNSS position onto the trajectory, a standard process of map-matching approaches, in this case, our interest is to select a probable region where the true position of the vehicle may be located. This process is described in the next section.

3.1.1 Candidates search space definition

For the LC INS/MAP, the candidate search space area is selected by using the latest INS navigated solution available after a TC GNSS/INS solution. By finding the closest point to the map, a candidate buffer area is obtained by selecting a pre-defined area in both directions of the closest point. This area selection is depicted in **Figure 5**.

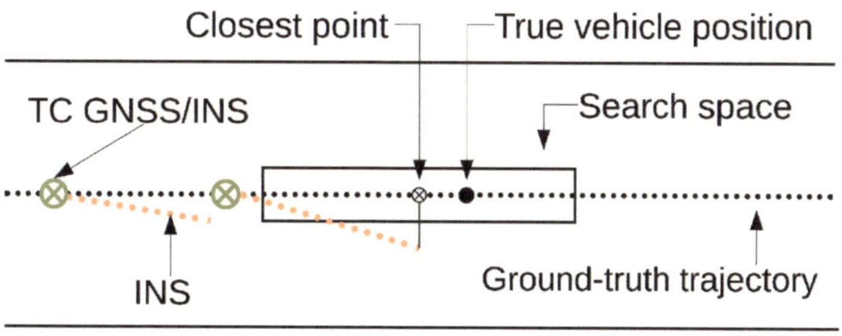

Figure 5.
LC INS/MAP searching space schema—top view.

This search space should match where the GNSS receiver is mounted on the vehicle. By accounting for this height difference we have a search space that will closely contain the true GNSS location, **Figure 6** illustrates this situation.

Once the search space, containing the most probable candidates, is defined, the best candidate selection can be performed. The next section explains this selection.

3.1.2 Candidate selection

The selection of the candidate is based on the Kalman filter residuals. The residuals, frequently called innovation term, tell us how well the model predicts the measurements, thus being a good measure of accuracy. Therefore, the smaller the residuals, the more accurate the estimate.

The measurement residuals (r_k) can be extracted from the second term multiplying the Kalman gain ($\boldsymbol{K}_k$) on the right-hand side of Eq. (27):

$$\boldsymbol{r}_k = \boldsymbol{z}_k - \hat{\boldsymbol{z}}_k^-, \tag{18}$$

where $\hat{\boldsymbol{z}}_k^- = \mathbf{h}(\boldsymbol{x}^0) - \boldsymbol{H}_k \boldsymbol{\delta}\hat{\boldsymbol{x}}_k^-$ is the predictive estimate of the measurement. $\boldsymbol{x}^0$ represents an initial information about the true state.

In the LC INS/MAP constraint, the measurement vector $\boldsymbol{z}_k$ is composed of position and velocity observations, as shown in Eq. (13).

As the search space, shown in **Figure 5**, gives a set of position candidates, they can be used as an approximation to the true states to generate the computed measurement

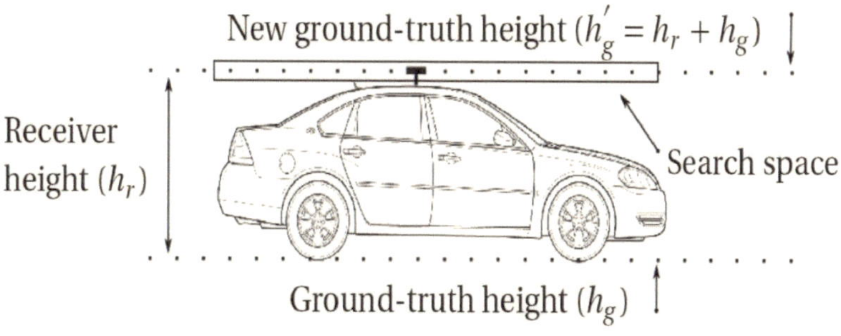

Figure 6.
Map-matching—side view.

estimate $\hat{z}$. Then, after the Kalman update step, each candidate generates a normalized (χ_c^2), as the following Equation:

$$\chi_c^2 = \left[\frac{\boldsymbol{r}_i^c}{\sigma_i^c}\right]^2. \tag{19}$$

where $i = 1, 2, \cdots, c$, with c being the number of candidates in the search space.

The best candidate is the one with the smallest residuals. To visualize this selection, an example showing six candidates in the search space is depicted in **Figure 7**. If there are no systematic errors in the measurements, candidate three would provide the smallest residual.

3.2 LKAS filter

3.2.1 Loosely coupled INS/MAP constraints

The LC INS/MAP filter is activated during GNSS outages, as shown in the full-system architecture 4. In this architecture, the MAP candidates are used to substitute the measurements in the LC Kalman measurement update.

Each candidate position, $\boldsymbol{r}^c$, from the search space, defined in **Figure 5**, substitute the LC measurement innovation, in Eq. (13), thus:

$$\delta \boldsymbol{z}_i = \begin{pmatrix} \boldsymbol{r}_i^c - \hat{\boldsymbol{r}}^{ins} \\ \boldsymbol{v}_i^c - \hat{\boldsymbol{v}}^{ins} \end{pmatrix}_k. \tag{20}$$

where $i = 1, 2, \cdots, c$, with c as the number of candidates in the search space.

For each LC measurement innovation generated in Eq. (20), a chi-square residual is computed for the best candidate selection with Eq. (19). After selected, the best candidate position is again substituted in Eq. (20), for a final LC measurement update.

An important piece of information is the velocity, $\boldsymbol{v}_i^c$, which is not actually available from the map. Since GNSS is not available, velocity information needs to be measured from another system. The INS could provide velocity, however, because of the accelerometer drift, after a few seconds, the INS derived velocities are not useful anymore [26]. A sensor that would provide such information would be wheel odometers.

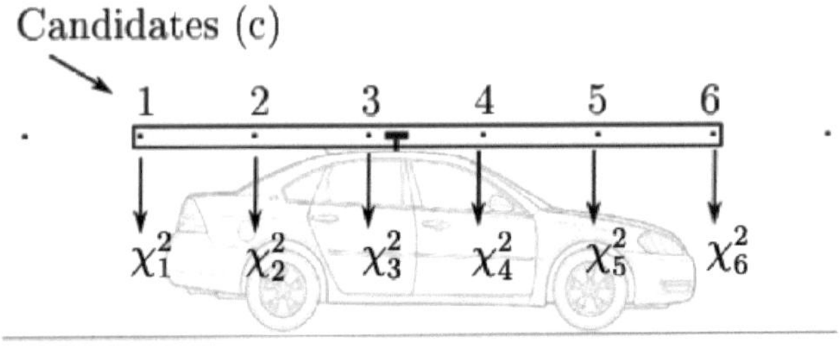

Figure 7.
Candidate space selection.

In this work, only a GNSS receiver and an IMU were used as instruments. Therefore, velocity information provided when GNSS was not available was obtained from the GNSS baseline solution, of the same data set, and is used as reference.

3.2.2 System solution transitions

The changes of the types of solutions will determine how the states are weighted. In the filter equations, key matrices to obtain smooth transition solutions are the state and measurement covariance, $\boldsymbol{P}_k$ and $\boldsymbol{R}_k$, respectively, and system covariance noise, $\boldsymbol{Q}_k$. The system noise covariance matrix reflects the sensor's noise, in this case, the IMU's accelerometers and gyroscopes. The IMU noise information can be seen in Appendix B. In Eq. (21), the covariance matrix is built with the IMU used in this experiment [26]:

$$\boldsymbol{Q}^e_{INS} = \begin{pmatrix} \mathbf{S}_{ra}\mathbf{I}_3 & 0_3 & 0_3 & 0_3 & 0_3 & 0 & 0 & 0 \\ 0_3 & \mathbf{S}_{rg}\mathbf{I}_3 & 0_3 & 0_3 & 0_3 & 0 & 0 & 0 \\ 0_3 & 0_3 & 0_3 & 0_3 & 0_3 & 0 & 0 & 0 \\ 0_3 & 0_3 & 0_3 & \mathbf{S}_{bad}\mathbf{I}_3 & 0_3 & 0 & 0 & 0 \\ 0_3 & 0_3 & 0_3 & 0_3 & \mathbf{S}_{bgd}\mathbf{I}_3 & 0 & 0 & 0 \\ 0_3 & 0_3 & 0_3 & 0_3 & 0_3 & S_{c\phi} & 0 & 0 \\ 0_3 & 0_3 & 0_3 & 0_3 & 0_3 & 0 & S_{cf} & 0 \\ 0_3 & 0_3 & 0_3 & 0_3 & 0_3 & 0 & 0 & q_T \end{pmatrix}$$
$$= \begin{pmatrix} 2.0409 & 0_3 & 0_3 & 0_3 & 0_3 & 0 & 0 & 0 \\ 0_3 & 5.3884^{-7} & 0_3 & 0_3 & 0_3 & 0 & 0 & 0 \\ 0_3 & 0_3 & 0_3 & 0_3 & 0_3 & 0 & 0 & 0 \\ 0_3 & 0_3 & 0_3 & 1.2769^{-12} & 0_3 & 0 & 0 & 0 \\ 0_3 & 0_3 & 0_3 & 0_3 & 2.3814^{-11} & 0 & 0 & 0 \\ 0_3 & 0_3 & 0_3 & 0_3 & 0_3 & 1 & 0 & 0 \\ 0_3 & 0_3 & 0_3 & 0_3 & 0_3 & 0 & 1 & 0 \\ 0_3 & 0_3 & 0_3 & 0_3 & 0_3 & 0 & 0 & 0.3 \end{pmatrix}. \tag{21}$$

where $\mathbf{S}_{ra}$ is in micro-g^2/h, $\mathbf{S}_{rg}$ in $\deg^2/h$, $\mathbf{S}_{bad}$ in m^2s^{-5}, and $\mathbf{S}_{bgd}$ in rad^2s^{-3}. The receiver clock frequency-drift and phase-drift are in m^2/s^3 and m^2/s, respectively. Troposphere standard deviation, q_T, is in meters.

Whenever a TC GNSS/INS solution is obtained, the measurement matrix values represent the GNSS carrier-phase and pseudorange observable errors in meters:

$$\boldsymbol{R}^{TC}_k = \begin{pmatrix} \sigma_P & 0 \\ 0 & \sigma_L \end{pmatrix} = \begin{pmatrix} 0.001 & 0 \\ 0 & 1 \end{pmatrix}. \tag{22}$$

When an LC INS/MAP is computed, a few modifications are done to accommodate the MAP observations. As the map tests its candidate space in the measurement update, in this step, we input the standard deviation of the map and and velocity in meters:

$$R_k^{LC} = \begin{pmatrix} \sigma_{r_{MAP}} & 0 \\ 0 & \sigma_{v_{MAP}} \end{pmatrix} = \begin{pmatrix} 10.0 & 0 \\ 0 & 0.1 \end{pmatrix}. \quad (23)$$

These values were derived after a manual tuning of the filter.

3.3 Lane-keeping accuracy limit

In this work, a lane accuracy limit is proposed based on the vehicle (v_w) and lane (l_w) dimensions. The relationship between the vehicle and lane widths is depicted in the **Figure 8**.

The accuracy criteria is derived from the average lane sizes where vehicle navigates and the vehicle lateral widths into the following equation:

$$\sigma_{lim} = \left(l_{w/2}\right) - \left(v_{w/2}\right). \quad (24)$$

For the real-scenario experiments, few lane width measurement samples were taken in Google Earth. In downtown, the average lane size was 3.75 m, and outside downtown, the average lane width was 4.10 m. Thus, the average lane size for the real-scenario experiment is $l_w = 3.93$ m. The vehicle used had a width of $v_w = 1.84$ m, thus the accuracy limit is $\sigma_{lim} = \pm 1.04$ m.

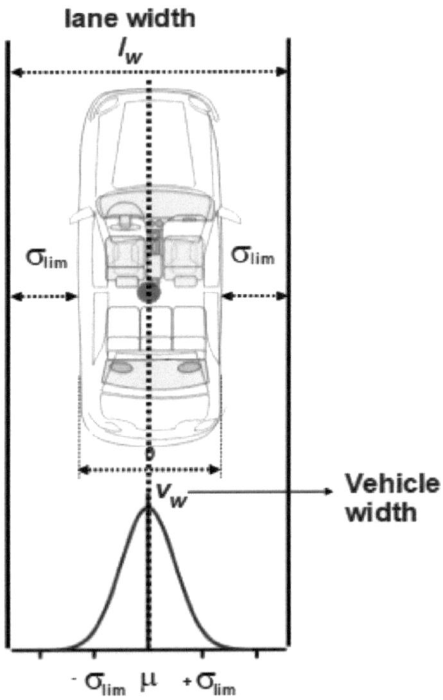

Figure 8.
Vehicle navigation threshold.

4. Real-world experiment

In this section, we present a real-world experiment conducted to test the proposed system shown in Section 3. The data required for the tests are presented in Section 4.1. In the next Section, 4.2, we show the results obtained along with the discussions.

4.1 Experiment setup

A GNSS and inertial dataset was collected onboard a vehicle driving in the streets of Fredericton, New Brunswick, Canada. The instruments used were a Septentrio Altus APS3G GNSS receiver, multi-constellation, and multi-frequency (L1, L2, and L5); and a tactical-grade KVH TG-6000 IMU. The GNSS receiver and IMU data rates were 1 and 100 Hz, respectively. The sensors were mounted aligned with the longitudinal axis of the vehicle, with the appropriate offsets accounted for during the processing.

The road map, composed with the location of the center of the streets, was created from the dataset with centimeter-accurate post-processing GNSS relative processing. A section of the road map is shown in **Figure 9**.

To test the proposed system, the driving test areas were selected where GNSS signal tracking was momentarily lost, with instances of complete signal interruption for approximately 20 seconds. For example, in **Figures 10-12**, the following areas were tested: walking bridge, road overpass, and streets with tall buildings, respectively.

Figure 9.
Road map of lanes obtained from GNSS relative processing.

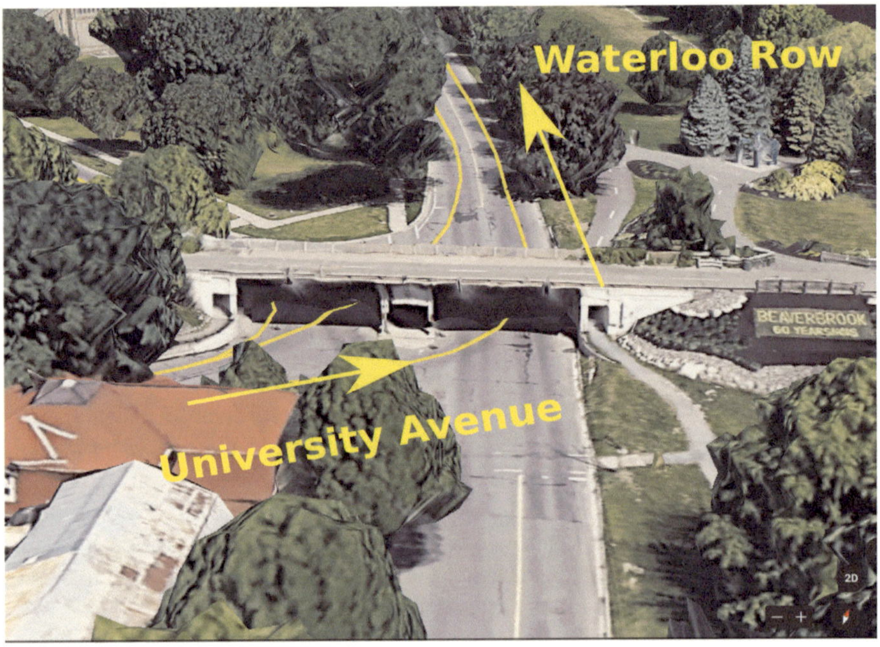

Figure 10.
Walking bridge.

Figure 11.
Road overpass.

Figure 12.
Street with tall buildings on both sides.

4.2 Results and discussions

This section presents an assessment of the proposed architecture, the TC GNSS/INS + LC/MAP, compared to a standard PPP (without convergence and re-convergences solutions) and TC GNSS/INS. The two GNSS/INS solutions are compared in terms of horizontal ground-track, in Section 4.2.1, accuracy analysis, adding PPP in analysis, in terms of cross and along-tracks, and up off-track differences in Section 4.2.2. In Section 4.2.3, a cross-track assessment based on the position accuracy criteria developed in Section 3.3, and finally, some solution challenges are described in Section 4.3.

The full-system solution will be referred to as TC + LC/MAP in the text and tables. For the figures, a separation of TC GNSS/INS and LC INS/MAP is made for a better visualization of the these solutions.

Knowing that PPP convergence and re-convergences greatly distort the accuracy of PPP solution, these periods were removed in order to have a fair comparison with the proposed architecture. Thus, the standard PPP solution in the accuracy analysis does not represent the real PPP accuracy for this trajectory.

4.2.1 Horizontal ground-track displacement

The horizontal comparisons in this section presents the behavior of the filters in places where the GNSS lost signal tracking. These places are the walking bridge, Forest Hill highway underpass, and downtown area with tall buildings.

The standard TC GNSS/INS is presented in green, with squares representing GNSS/INS integration and lines representing the INS solution. The proposed TC + LC/MAP solution is represented in gold and red. The integrated TC GNSS/INS is represented with gold circles, and integrated LC INS/MAP is shown with a red star, while gold lines represent the INS solution.

The first obstruction happens on the walking bridge, as shown in **Figure 10**. The walking bridge is a short overpass that obstructs the GNSS signal for a brief moment. From the solution horizontal plot in 13, the TC GNSS/INS solution goes off the driving path as GNSS is lost. This behavior happens because the IMU does not have a reference to correct its drift. For the TC+LC/MAP, a position solution is available from the LC INS/MAP filter, thus bridging the GNSS signal outage during the obstruction **Figure 13**.

At the Forest Hill highway underpass, as depicted in **Figure 11**, the longest complete GNSS signal interruption occurs in the experiment. Similar to the walking bridge, the TC GNSS/INS solution drifts off the trajectory, whereas the TC + LC/MAP has a continuous solution with the LC INS/MAP, as shown in **Figure 14**.

In the downtown area, a sequence of tall buildings on both sides of the street caused difficulties in GNSS signal tracking for about 20 seconds. This area is presented in **Figure 12**. During these blocks, the GNSS receiver suffered consecutive obstructions as seen in **Figure 15**, causing the TC GNSS/INS solution to drift due to having only INS solution available, as can be seen by a long green line disappearing out of plot. Once more, the proposed architecture provides a LC INS/MAP solution throughout the obstructions, effectively bridging the GNSS outages.

4.2.2 Off-track accuracy analysis

The cross and along-tracks differences between the ground-truth to the solutions are compared in histograms plots in **Figures 16** and **17**.

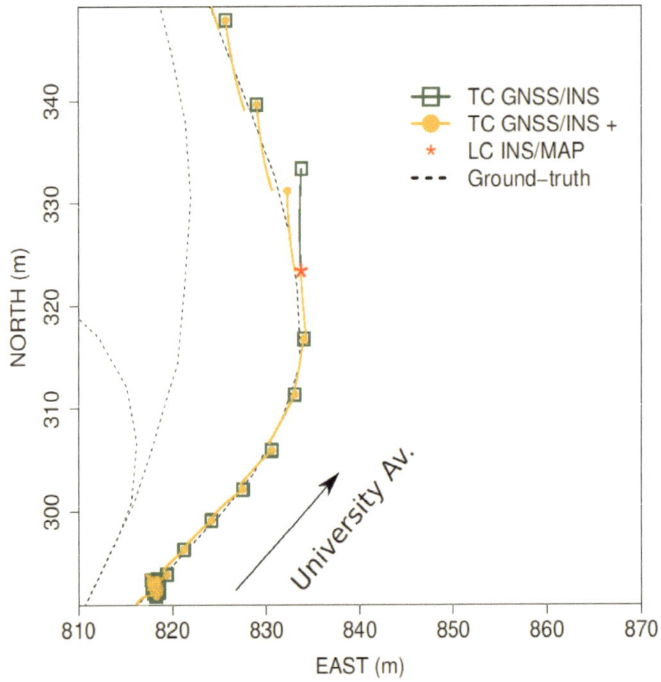

Figure 13.
Walking bridge obstructions.

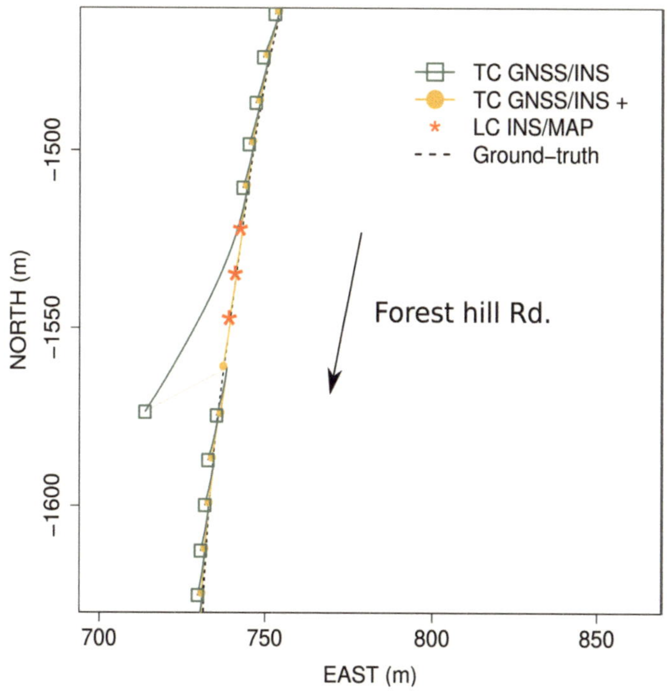

Figure 14.
Forest Hill underpass.

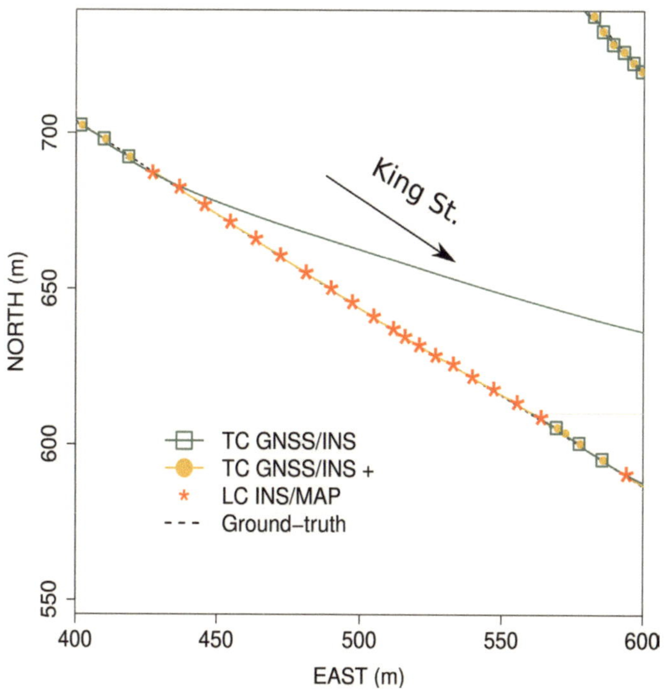

Figure 15.
Downtown area—king street tall buildings.

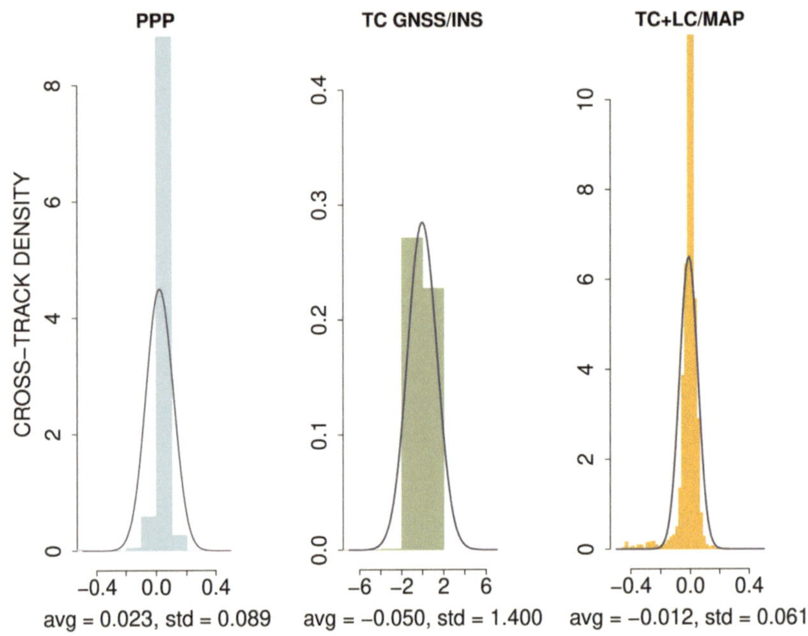

Figure 16.
Cross-track histograms.

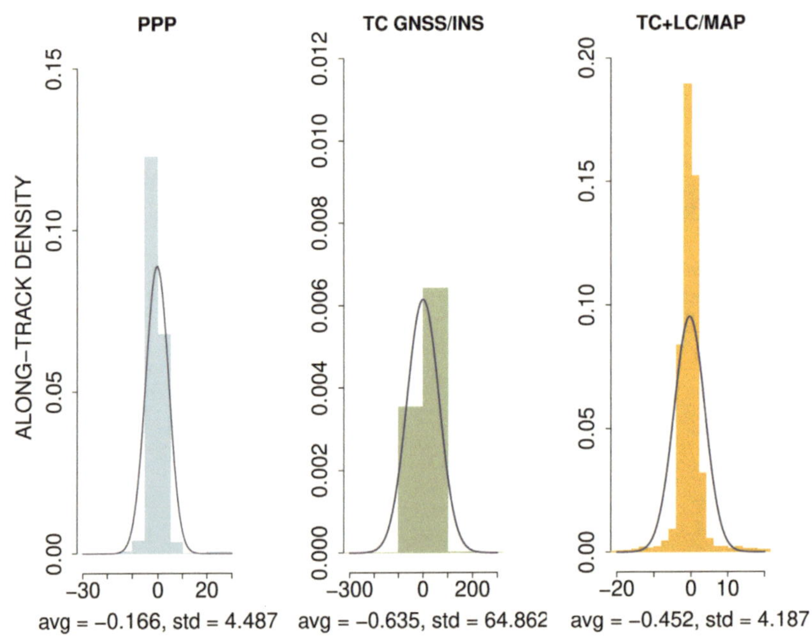

Figure 17.
Along-track histograms.

The TC GNSS/INS histograms presented the larger tails with a low frequency of samples localized around zero. This shows that a large amount the differences were off from the lane center, meaning that it was neither accurate nor precise. This behavior,

	Cross-track		Along-track	
	avg (m)	**std (m)**	**avg (m)**	**std (m)**
PPP	0.02	0.09	−0.17	4.49
TC GNSS/INS	−0.05	1.40	−0.63	64.86
TC + LC/MAP	−0.01	0.06	−0.45	4.19

Table 1.
Cross and along-tracks off-track statistics.

presented in plots from Section 4.2.1, happens during long periods of GNSS obstructions, where the INS drifts several meters off the trajectory.

TC + LC/MAP presented better performances than PPP in cross-track, with a higher frequency of samples around zero and smaller spread, in other words, more accurate and precise, as can be seen in **Figure 16**. For the along-track, the contrary is true, PPP achieved a better accuracy and precision, 17.

Table 1 summarizes the cross-track and along-track-off-track mean and standard deviations. In terms of cross-track performance, the proposed system outperformed the PPP by 1 cm in average and by 3 cm in standard deviation. However, in along-track measurements, the full-system did worse than PPP in terms of average by 28 cm and outperformed in standard deviation by 30 cm. As mentioned previously, TC GNSS/INS had the worse performances in cross and along-tracks due to INS solution drifts.

4.2.3 Cross-track accuracy criteria evaluation

As the positioning accuracy criteria developed in Section 3.3 consider lateral lane offsets, only cross-track time series is analyzed in this section.

Figure 18 shows the cross-track for the three solutions. As seen in Section 3.3, the lane accuracy limit for analyzing cross-track is ± 1.04 m, represented by the dashed red lines. **Table 2** shows the statistics about the cross-tracking series of **Figure 18**.

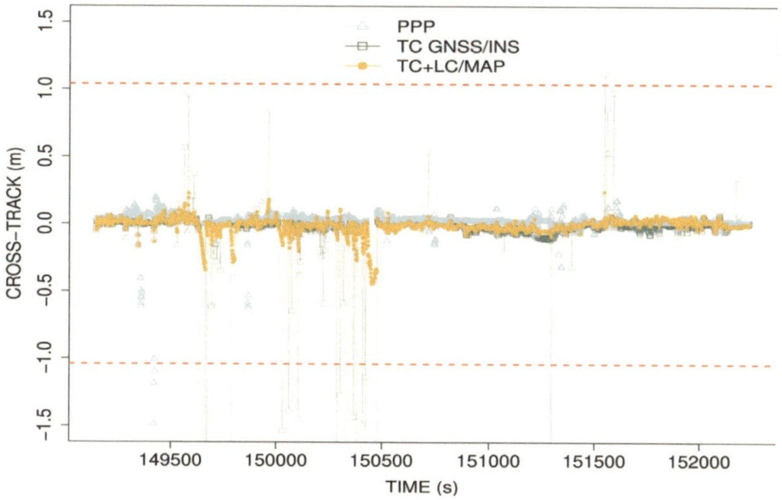

Figure 18.
Cross-track PPP, TC GNSS/INS, and TC + LC/MAP.

	avg (m)	std (m)	95% C.I.	
			Lower bound	Upper bound
PPP	0.02	0.09	−0.15	0.20
TC GNSS/INS	−0.05	1.40	−2.79	2.69
TC + LC/MAP	−0.01	0.06	−0.13	0.11

Table 2.
Cross-track statistics.

In summary, the TC + LC/MAP solutions were more accurate and precise than the others, and, with 95% of confidence level, only TC GNSS/INS did not achieve the lane accuracy limit established.

PPP and TC + LC/MAP solutions had respective biases of 2 and 1 cm, and standard deviations of 13 and 11 cm. The TC GNSS/INS solution had a bias of 5 with 140 cm standard deviation.

Based on a more strict position accuracy limit for lane-level, presented in Section 3.3, the TC + LC/MAP was 3 and 1 cm away from achieving the active control level (0.1 m) when looking at the 95% lower and upper bounds of the cross-track confidence interval series: −0.13 and 0.11, respectively.

4.3 Challenges

This section will explore the reasons the proposed system along-track did worse than the PPP solution, as presented in **Table 1**.

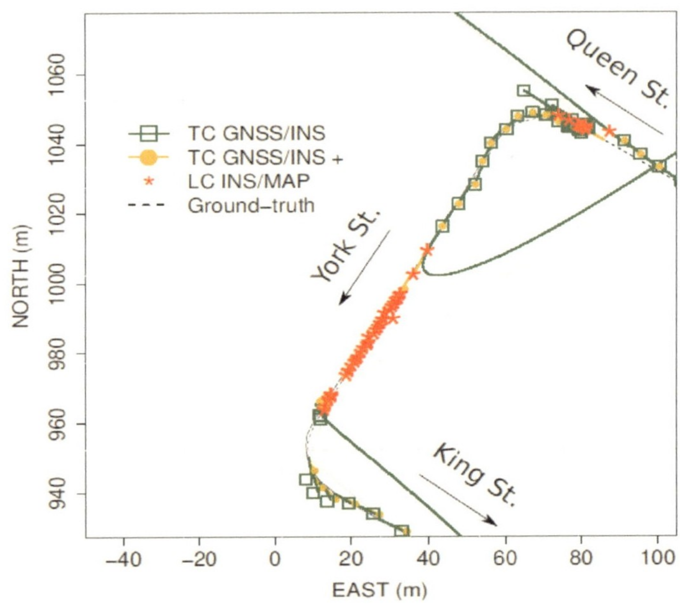

Figure 19.
Downtown area first run—Queen/York and York/King street corners.

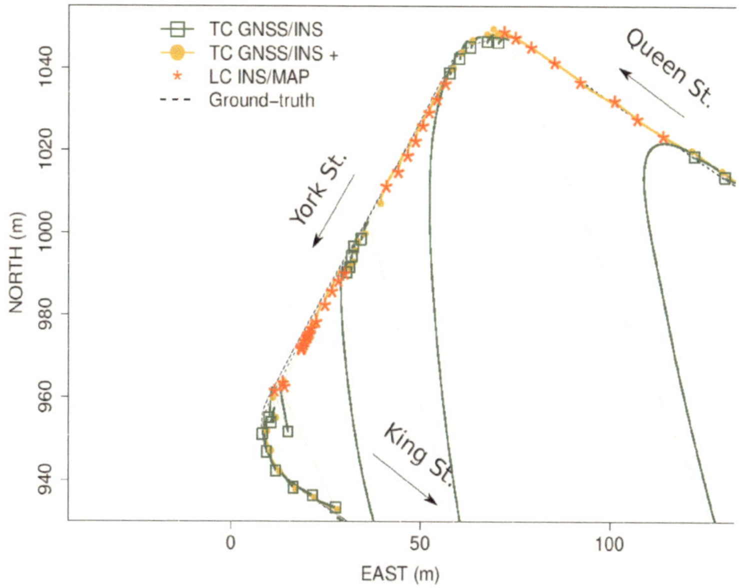

Figure 20.
Downtown area third run—Queen/York and York/King street corners.

This region is located in an area covered by trees on both sides of the street. The solutions analyzed are the first and third runs in downtown, shown in **Figures 19** and **20**, respectively.

A few gaps are observed after LC + INS solutions from the proposed system. The reason of this behavior is the lack of reliable velocity information.

Integration spikes are also observed in the TC GNSS/INS solutions, represented by the green boxes in **Figures 19** and **20**. According to the filter design, in GNSS quality control, integration would only happen when the GNSS solution is reliable. Therefore, in these cases, the probable source of such behavior is signal multipath, which is still a major challenge in GNSS positioning research.

5. Conclusions

This work proposed a satellite and INS with map constraints integration filter architecture. The major contribution of this architecture was to demonstrate the possibility of lane-level vehicle navigation using a simple set of sensors: a GNSS receiver, an IMU, and a map of the lanes. A TC GNSS/INS with LC INS/MAP architecture was developed.

In the real-world experiments, the system was exposed to complex urban situations. The system was able to correctly maintain solutions centered during long GNSS outages of 20 seconds with the LC INS/MAP filter. The elimination of INS drift, a common behavior when there is no information to calibrate the IMU. A position accuracy was achieved in cross-track, with 95% confidence, the full-system cross-track lower and upper bounds were 3 and 1 cm away, respectively, from achieving the active control limit level (0.1 m) for ITLBS.

Along with current lane-keeping assistant systems, this new technique could be a strong addition to vision approaches due to its independence of lighting conditions.

This system has potential for global operation since the proposed system is not restricted in terms of the area of operation, as current approaches are based on machine learning algorithms, which normally depend on limited training data of a specific region.

One of the challenges for such a system to operate worldwide is the mapping coverage. In recent years, private companies have increased their interest in road mapping. However, for a long-term and larger-scale solution, an interesting approach to solve the problem would be the involvement of the public sector to provide such maps and be responsible for updating and maintaining them, as this is a matter of public road safety.

Future improvements could include the addition of other sensors, for instance, wheel odometers that could provide velocity information for longer GNSS obstructions. Such an addition could also enable indoor navigation.

Acknowledgements

This work was funded by the Brazilian National Council for Scientific Technological Development (CNPq) through the Sciences Without Borders (SwB) program.

Appendix A: discrete kalman filter

From the system and measurement continuous models respectively presented in **Tables 1** and **3**, the discrete Kalman filter system and measurement equations are represented in the following form, respectively:

$$\begin{aligned} \boldsymbol{x}_k^- &= \boldsymbol{F}_k \boldsymbol{x}_{k-1}^+ + \boldsymbol{v}_{k-1} \\ \boldsymbol{z}_k &= \boldsymbol{H}_k \boldsymbol{x}_k^- + \boldsymbol{w}_k. \end{aligned} \tag{25}$$

The Kalman filter is a recursive process with the system prediction and measurement update steps. In the system prediction step, the state and error covariance are estimated from previous time-step:

$$\begin{aligned} \hat{\boldsymbol{x}}_k^- &= \boldsymbol{\Phi}_{k-1} \hat{\boldsymbol{x}}_{k+1}^+ \\ \boldsymbol{P}_k^- &= \boldsymbol{\Phi}_{k-1} \boldsymbol{P}_{k-1}^+ \boldsymbol{\Phi}_{k-1}^T + \boldsymbol{Q}_{k-1}, \end{aligned} \tag{26}$$

where $\boldsymbol{P}_k$ is the state error covariance, and the transition matrix $bmUpphi_{k-1}$ is computed as a power-series expansion of the system matrix $\boldsymbol{F}_{k-1}$.

As new measurements are available, the predicted state and covariances are corrected in the measurement update step:

$$\begin{aligned} \hat{\boldsymbol{x}}_k^+ &= \hat{\boldsymbol{x}}_k^- + \boldsymbol{K}_k \left(\boldsymbol{z}_k - \boldsymbol{H}_k \hat{\boldsymbol{x}}_k^- \right) \\ \boldsymbol{P}_k^+ &= (\boldsymbol{I} - \boldsymbol{K}_k \boldsymbol{H}_k) \boldsymbol{P}_k^-, \end{aligned} \tag{27}$$

where $\boldsymbol{K}_k = \boldsymbol{P}_k^- \boldsymbol{H}_k^T \left(\boldsymbol{H}_k \boldsymbol{P}_k^- \boldsymbol{H}_k^T + \boldsymbol{R}_k \right)^{-1}$ is the Kalman gain. The Kalman gain weights the state estimation according to the system prediction and measurements update quality.

Appendix B: KVH TG-600 tactical INS noise

See **Tables 3** and **4**.

	$\mathbf{m/s^2\sqrt{Hz}}$
PSD	0.000014
Bias	0.000001

Table 3.
KVH accelerometers noise.

	$\mathbf{rad/s^2\sqrt{Hz}}$
PSD	0.000769
Bias	0.000005

Table 4.
KVH gyroscopes noise.

Author details

Emerson Pereira Cavalheri[1†] and Marcelo Carvalho dos Santos[2*†]

1 Volkswagen of Brazil, São Bernardo do Campo, Brazil

2 University of New Brunswick, Fredericton, Canada

*Address all correspondence to: msantos@unb.ca

† These authors contributed equally.

References

[1] World Health Statistics. Monitoring Health for the SDGs Sustainable Development Goals. Geneva, Switzerland: World Health Organization; Available from: https:// www.who.int/data/gho/data/themes/ topics/topic-details/GHO/road-traffic-mortality; 2020 [Accessed: May 3, 2024]

[2] Chao E. Automated Driving Systems 2.0: A Vision for Safety. National Highway Traffic Safety Administration and Others. Vol. 812. Washington, DC: US Department of Transportation, DOT HS; 2017. p. 442

[3] Katriniok A. Optimal Vehicle Dynamics Control and State Estimation for a Low-cost GNSS-based Collision Avoidance System. Germany: Hochschulbibliothek der Rheinisch-Westfälischen Technischen Hochschule Aachen; 2014

[4] Souza JR, Sales DO, Shinzato PY, Osório FS, Wolf DF. Template-based autonomous navigation and obstacle avoidance in urban environments. ACM SIGAPP Applied Computing Review. 2011;**11**(4):49-59

[5] Li Q, Chen L, Li M, Shaw S-L, Nüchter A. A sensor-fusion drivable-region and lane-detection system for autonomous vehicle navigation in challenging road scenarios. IEEE Transactions on Vehicular Technology. 2013;**63**(2):540-555

[6] Hillel AB, Lerner R, Levi D, Raz G. Recent progress in road and lane detection: A survey. Machine Vision and Applications. 2014;**25**(3):727-745

[7] Chokshi N. Tesla Autopilot System Found Probably at Fault in 2018 Crash. New York Times; 2020. Available from: https://www.nytimes.com/2020/02/25/business/tesla-autopilot-ntsb.html. 2020 [Accessed: June 9, 2024]

[8] Cellan-Jones R. Uberś Self-driving Operator Charged Over Fatal Crash. BBC Electronic News; 2020. Available from: https://www.bbc.com/news/technology-54175359 [Accessed: June 9, 2024]

[9] Schmidt GT, Phillips RE. INS/GPS integration architecture performance comparisons. Electronic NATO RTO Lecture Series, RTO-EN-SET-116, Low-Cost Navigation Sensors and Integration Technology U.S Government Publishing Office. 2011. Available from: https://www.sto.nato.int/publications/STO%20Educational%20Notes/Forms/Document%20Set%20View.aspx?RootFolder=%2Fpublications%2FSTO%20Educational%20Notes%2FRTO%2DEN%2DSET%2D116%2D2011&FolderCTID=0x0120D5200078F9E87043356C409A0D30823AFA16F60300099FA443AE6E08499A57A0FBE0134F20&View=%7BB927897E-9DC2-4392-AA25-598B0C04B48E%7D. [Accessed: June 11, 2024]

[10] Bishop R. Intelligent Vehicle Technology and Trends. London: Artech House Publishers; 2005

[11] Taylor G, Blewitt G. Intelligent Positioning: GIS-GPS Unification. ISBN: 978-0-470-03566-5. Chichester, England: Wiley Online Library; 2006

[12] Enhancement of global vehicle localization using navigable road maps and dead-reckoning. In: Position, Location and Navigation Symposium. Monterey, California, United States: IEEE; 2008. pp. 1286-1291

[13] Groves PD, Jiang Z,Wang L, Ziebart MK. Intelligent Urban Positioning Using

Multi-Constellation GNSS with 3D Mapping and NLOS Signal Detection. In: Proceedings of the 25th international technical meeting of the satellite division of the Institute of Navigation (ION GNSS 2012) 2012. pp. 458-472

[14] Groves PD. It's Time for 3D Mapping-aided GNSS. London: Inside GNSS Magazine; Available from: https://www.insidegnss.com/auto/sepoct16-GROVES.pdf. 2016. pp. 50-56

[15] Groves P.D. Principles of GNSS Inertial, and Multisensor Integrated Navigation Systems. Boston, London: Artech House; 2013

[16] Xu G, Xu Y. GPS. Berlin, Heidelberg: Springer-Verlag; 2007

[17] Hofmann-Wellenhof B, Lichtenegger H, Wasle E. GNSS—Global Navigation Satellite Systems: GPS, GLONASS, Galileo, and More. Mörlenbach, Germany: Springer Science and Business Media; 2007

[18] Teunissen P, Montenbruck O. Springer Handbook of Global Navigation Satellite Systems. Cham, Switzerland: Springer; 2017

[19] Tao C, Li R, Chapman MA. Automatic reconstruction of road centerlines from mobile mapping image sequences. Photogrammetric Engineering and Remote Sensing. 1998; **64**:7

[20] Tao CV. Mobile mapping technology for road network data acquisition. Journal of Geospatial Engineering. 2000; **2**(2):1-14 . The Hong Kong Institution of Engineering surveyors

[21] Elghazaly G, Frank R, Harvey S, Safko S. High-definition maps: Comprehensive survey, challenges and future perspectives. IEEE Open Journal of Intelligent Transportation Systems. 2023;4:527-550

[22] Strijbosch W. Safe autonomous driving with high-definition maps. ATZ Worldwide. 2018;**120**(11):28-33

[23] Petovello MG. Real-Time Integration of a Tactical-Grade IMU and GPS for High-Accuracy Positioning and Navigation. Canada: University of Calgary Library; Available from: https://prism.ucalgary.ca/handle/1880/42707. [Accessed: June 11, 2024]

[24] Quddus MA, Ochieng WY, Zhao L, Noland RB. A general map matching algorithm for transport telematics applications. GPS Solutions. 2003;7(3): 157-167

[25] Quddus MA, Ochieng WY, Noland RB. Current map-matching algorithms for transport applications: State-of-the art and future research directions. Transportation Research Part C: Emerging Technologies. 2007;**15**(5): 312-328

[26] El-Sheimy N, Hou H, Niu X. Analysis and modeling of inertial sensors using Allan variance. IEEE Transactions on Instrumentation and Measurement. 2007;**57**(1):140-149